Guide to

Spectroscopic Identification

of

Organic Compounds

Guide to

Spectroscopic Identification of

Organic Compounds

Karen Feinstein

CRC Press
Taylor & Francis Group
Boca Raton London New York

CRC Press is an imprint of the
Taylor & Francis Group, an **informa** business

Acknowledgments

I would like to thank Dr. Earl Baker, Christopher Birdsall, Jim Brody, Barbara Caras, Harvey Feinstein, and Evelyn Sternberg for their special efforts in producing the *Guide to Spectroscopic Identification of Organic Compounds*.

The mass spectra and ^{13}C NMR spectra were drawn by Chris and Barbara using available data from the *Handbook of Data on Organic Compounds (HODOC) database*.

Permission to use the IR spectra and ^{1}H NMR spectra has been graciously allowed by Sadtler Research Laboratories. "Permission for the publication herein of Sadtler Standard Spectra® has been granted, and all rights are reserved, by Sadtler Research Laboratories, Division of Bio-Rad Laboratories, Inc."

^{13}C NMR chemical shifts predictions were determined using SoftShell International's *Carbon-13 NMR Module*. This module was developed by Dr. Erno Pretsch and Andras Furst, Department of Organic Chemistry, ETH Zurich. The standard deviation for the predicated chemical shifts is 5.5 ppm.

CRC Press
Taylor & Francis Group
6000 Broken Sound Parkway NW, Suite 300
Boca Raton, FL 33487-2742

© 1995 by Taylor & Francis Group, LLC
CRC Press is an imprint of Taylor & Francis Group, an Informa business

No claim to original U.S. Government works

ISBN-13: 978-0-8493-9448-5 (hbk)

Visit the Taylor & Francis Web site at
http://www.taylorandfrancis.com

and the CRC Press Web site at
http://www.crcpress.com

Library of Congress Cataloging-in-Publication Data
Catalog record is available from the Library of Congress.

Preface

Instruments including mass (MS), infrared (IR), nuclear magnetic resonance (NMR), and ultraviolet (UV) spectrophotometers have greatly enhanced the traditional techniques of structure determination. At this time an investigator can elucidate the structure of a molecule wholly from the complementary spectra obtained from using these devices. The *Guide to Spectroscopic Identification of Organic Compounds* is intended to be a helpful and convenient tool for the analyst in interpreting organic spectra. It may serve as a companion piece to any organic textbook or a spectroscopy reference; its size will allow practitioners to carry it along when other tools might be cumbersome or expensive. A general problem-solving algorithm is presented to provide direction to the analyst who must piece together the discrete parts so that one cohesive structure results.

MS spectra allow for precise determination of molecular weights. Characteristic fragment ions point to specific functional groups. IR spectra confirm or reject the presence of certain functionalities. ^{1}H NMR spectra capitalize on the fact that different protons in different chemical environments experience differential shielding and therefore, have different chemical shifts. ^{13}C NMR is a useful extension to proton spectra; different carbon atoms absorb in a distinct range. UV spectra are particularly helpful in identifying unsaturation or the presence of heteroatoms having non-bonding electrons.

Representative compounds are analyzed using the above general procedure. The examples are solved using the standard but flexible method outlined. Predominant features in one spectrum and differential given information dictate subtle and necessary deviations in the explanation pattern. To illustrate, the classic heptet peak in an ^{1}H NMR spectrum drives the analysis. That feature becomes the prime starting point. Solutions are eclectic, ranging from simple and straightforward to complex. Compounds and their derivatives, structural isomers, straight chain, and aromatics are selected to illustrate didactic concepts. Practice problems are provided as well.

An additional feature of this *Guide* is the data extracted from the *Handbook of Data on Organic Compounds* (HODOC) database. The analyst is provided a picture of the relationship of structure to physical properties and spectral features. Properties such as boiling point, melting point, density, and refractive index are helpful in confirming or repudiating possible structural candidates. Identifiers including CAS Registry Numbers and Beilstein References can be used as starting points for research into preparation and reaction information.

A NOTE TO STUDENTS

Students often learn in an optimal way when all the tools they need are conveniently placed in clear view and at arm's reach. I recommend that you photocopy spectroscopic aids (particularly the "Molecular Formula Compilation Table", diagnostic fragment ions in the MS spectrum, diagnostic IR absorption frequencies, approximate chemical shifts in ^{1}H NMR, and UV absorptions of chromophores).

- Be mindful of the "helpful hints" page of the *Guide* and approach each unknown with the problem-solving algorithm tool.

- Make certain your environment is well lit.

- Cover the solution of the problem. Try to solve an unknown problem solution on your own.

- After you have arrived at a tentative structure or have decided you have been "stuck" long enough, refer to the explanation. The specific path you take in your solution need not match the author's. If your approach works, that's what counts. Consider the author's reasoning as an alternative.

- Your input is important to me. I am hopeful that this *Guide* is a truly helpful tool. Let me know if there is some information that would enhance this book.

Good luck!

Karen

Table of Contents

Appendices

Indices

The General Problem-Solving Approach

General Problem-Solving Algorithm

Students may use this tool to elucidate organic structures based on the correlation of information from the MS, IR, ^1H NMR, ^{13}C NMR, and UV spectra and/or spectral data.

Given: A combination of MS, IR, ^1H NMR, ^{13}C NMR, UV spectra or major peaks or chemical shifts of an unknown compound.

Find: ? Identity.

Solution:

1. Eyeball the four complementary spectra or combination of major shifts or peaks, looking for any predominant features.

2. Determine the molecular weight of the unknown by inspection of the molecular ion in the MS (numerically equivalent to the m/e for dominant isotopic species).

3. Establish the presence or absence of heteroatoms other than nitrogen or oxygen, hereafter referred to as remarkable heteroatoms, using the M+1 and M+2 ions.

4. Subtract the masses of elements other than carbon, hydrogen, oxygen, nitrogen from the molecular weight.

5. Use the "Molecular Formula Compilation Table" in the *Guide*, a condensed version of Beynon's Table, to list the possible and likely candidates.

6. Calculate the level of unsaturation (LU) using the equation:

$$LU = C+1-H/2-X/2+N/2.$$

7. Inspect the IR spectrum to confirm or negate the presence of characteristic absorptions indicating particular functional groups.

8. Subtract the number of atoms from any functional group identified.

9. Analyze the ^1H NMR and ^{13}C NMR spectra to determine how many different types of protons and carbon atoms are present. Notice diagnostic chemical shifts. Integrate peaks to give the number of relative protons attached to each carbon atom under consideration.

10. Confirm unsaturation and the presence of heteroatoms having non-bonding electrons by checking absorption in the UV range.

11. Assemble the data.

12. Confirm the best fit.

Helpful Hints

- Use the obvious features of one spectrum to elucidate the subtleties of another spectrum.

- Resist the temptation to interpret every peak or chemical shift in a given spectrum.

MS Spectrum

- When using the "Molecular Formula Compilation Table", notice that molecular formulas that contain oxygen have comparatively large M+2 peaks due to the natural abundance of ^{18}O.

- A large isotope contribution to M+2 indicates the presence of heteroatoms such as sulfur (4.4% of molecular ion), chlorine (1/3 as large as molecular ion), or bromine (equal in intensity to the molecular ion).

- A small M+1 relative to the molecular ion indicates the presence of iodine, fluorine, or phosphorus.

- Fragments are favored that result from elimination of small stable molecules such as:

H_2O	18
CO	28
NH_3	17
HCN	27
RSH	33+R
ROH	17+R

- Cleavage is favored at branched carbon atoms. The largest substituent at a branch is eliminated most readily as a radical.

- Carbon-carbon bonds adjacent to heteroatoms are frequently cleaved, leaving the charge on the fragment containing non-bonding electrons of the heteroatom due to resonance stabilization.

- A straight chain molecule has peaks at predictable increments of 14.

4

IR Spectrum

- This spectrum definitively proves or disproves the presence of certain functional groups. Look for distinctive absorptions confirming or negating typical functional group, e.g. O-H stretching vibrations at 3200—3600 cm^{-1}.

- A positive identification of the unknown can be made by a comparison of its IR spectrum with that of the known suspected compound. The fingerprint regions of the two should be a perfect match. This region is unique for each compound.

^1H NMR and ^{13}C NMR Spectrum

- The number of chemical shifts in the ^1H NMR indicates the number of different protons in the molecules. A difference between the number of hydrogen atoms in the molecular formula and the number of chemical shifts in the spectrum indicates symmetry in the molecule.

- Be aware of typical chemical shifts indicating the presence of characteristic functional groups, e.g. aldehydic protons absorb at 9—10 ppm.

- The number of chemical shifts in the ^{13}C NMR spectrum indicates the number of different carbon atoms in the structure of the molecules. Subsequently, a difference between the number of carbon atoms present in the molecular formula and the number ^{13}C NMR chemical shifts in the spectrum indicates symmetry in the molecule.

UV Spectrum

- Because the energy required for $\sigma \rightarrow \sigma^*$ transitions is very high, saturated hydrocarbons do not show absorption in the ordinary UV region.

- Compounds containing non-bonding electrons (due to the presence of oxygen, nitrogen, sulfur, or a halogen) require less energy because absorptions are due to $\eta \rightarrow \sigma^*$ transitions. These transitions do appear in UV spectrum.

- Those molecules with unsaturation sites exhibit transitions to antibonding π * orbital, specifically $\pi \rightarrow \pi^*$ transitions. These transformations are low energy ones and thus occur at longer wavelengths, but still in the UV region.

- Conjugation in the molecule increases the wavelength of absorption. Colored compounds have extensive conjugation which causes them to absorb in the visible region, thus producing the observed hue.

Sample Compound: Heptanal

GIVEN: IR spectrum, m/e, MS spectrum, ¹H NMR chemical shifts
FIND: Identity of Compound A

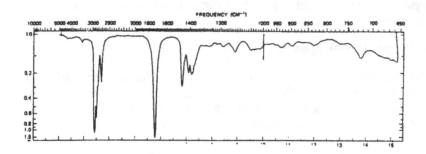

Figure 1. IR compound A. (With permission from
Bio-Rad/Sadtler Division.)

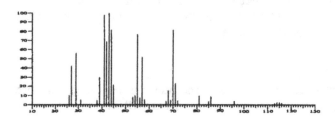

Figure 2. MS compound A.

MS Reference: **NIST 61961**
MS Peaks (Intensities): 43(100) 41(97) 70(82) 44(82) 55(76)42(68) 29(56)
  57(52) 27(43) 39(30)
m/e for Dominant Isotopic Species: 114.10
IR Reference: **COB 5731**
IR Peaks [cm⁻¹]: 2940 2850 2710 1730 1470 1380 1140 950 720
¹H NMR Reference: **SAD 9267**
¹H NMR Shifts [ppm]: 0.9 1.3 2.4 4.8 9.7 CCl₄

6

SOLUTION

This compound analysis illustrates the deductive process used to narrow the list of candidates from the "Molecular Formula Compilation Table".

1. The IR spectrum contains a characteristic carbonyl absorption at 1730 cm^{-1}. The base peak of the MS is 43; peaks at increments of 14 (m/e = 29, 43, 57) are present, indicating a straight chain configuration. A chemical shift of 9.7 ppm in the ^{1}H NMR spectrum is clearly that of an aldehydic proton. The number of peaks indicates a minimum of four different hydrogen atoms.

2. An even molecular weight of 114 indicates zero or an even number of nitrogen atoms.

3. Inspection of the M+1 and M+2 peaks point to the absence of remarkable heteroatoms.

4. Possible candidates from the table are: $C_4H_2O_4$, $C_4H_6N_2O_2$ (UV does not indicate the degree of unsaturation necessary in the first two candidates), $C_4H_{10}N_4$ (can't be this one because there must be an aldehyde), $C_5H_6O_3$ (too unsaturated), $C_5H_{10}N_2O$, $C_6H_{12}O_2$, $C_7H_{14}O$.

5. The last three candidates are most likely. Their LU's are calculated.

   (1) LU = 5+1-10/2+2/2=2   This prospect is eliminated because the presence of nitrogen is not indicated. (2) LU = 6+1-12/2=1 or (3) LU = 7+1-14/2=1 These remain viable.

6. Subtraction of the known functional group yields $C_6H_{12}O_2$-CHO=$C_5H_{11}O$ The IR spectrum rules out the possibility that the additional oxygen atom is part of an alcohol or carboxylic acid. Then the best fit is $C_7H_{14}O$-CHO=C_6H_{13}.

CONFIRMATION

CAS Index Name: Heptanal
Molecular Formula: $C_7H_{14}O$
Line Formula: $CH_3(CH_2)_5CHO$
CAS Registry Number: 111-71-7
Beilstein Reference: 1[4], 3314
Molecular Weight: 114.18

Melting Point [°C]: -43.3
Solubility: H_2O 2; al 5; eth 5; ctc 2
Refractive Index: 1.4113[20]
Density [g/cm³]: 0.8495[20]
Boiling Point [°C]: 152.8

GIVEN: IR spectrum, ¹H NMR spectrum, MS peaks, m/e
FIND: Identity of Compound B

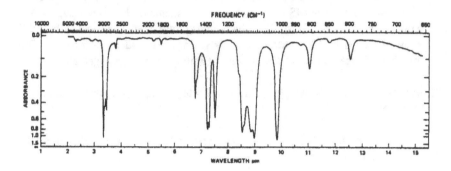

Figure 3. IR compound B. (With permission from
Bio-Rad/Sadtler Division.)

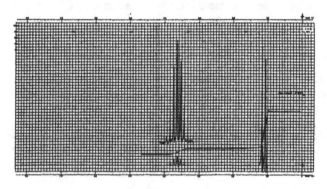

Figure 4. ¹H NMR compound B. (With permission from
Bio-Rad/Sadtler Division.)

MS Reference: **NIST 61391 WILEY 1125**
MS Peaks (Intensities): 45(100) 43(39) 87(15) 41(12) 59(10) 27(8) 39(4)
    69(3) 42(3) 31(3)
m/e for Dominant Isotopic Species: 102.10
IR Reference: μ **SADP 102**
IR Peaks [cm⁻¹]: 2940 2860 2630 1470 1390 1330 1180 1110 1020 910
    790
¹H NMR Reference: **SAD 7258**
¹H NMR Shifts [ppm]: 1.0 3.5 CCl₄

SOLUTION

This compound exemplifies the major role that the ^{1}H NMR may play in identifying a compound. The classic heptet and doublet in a 6:1 proton ratio indicate an isopropyl group.

1. The isopropyl group is apparent from the ^{1}H NMR. IR inspection eliminates NH, OH, carbonyl, alkene or alkyne structural units.

2. The molecular weight is 102 as indicated by the m/e.

3. M+1 and M+2 peaks preclude halogen or sulfur in the molecule.

4. Candidates from the table that do not exhibit unsaturation to any degree:
   $C_5H_{10}O_2$ and $C_6H_{14}O$

5. LU = 5+1-10/2=1, 6+1-14/2=0

6. The molecular formula $C_6H_{14}O$ works very well. Isopropyl ether fits due to its symmetry and the pattern of the ^{1}H NMR. Further confirmation comes in the form of the base peak of the MS, CH_3CHOH^+, which resulted from double cleavage with rearrangement of the hydrogen atom. Another prominent peak at m/e = 43 for C_3H_7 is due to the C-O cleavage with retention of the charge on the alkyl moiety. The peak at m/e = 87 is the result of the loss of a methyl group. The ether is further substantiated by the C-O-C IR absorption at 1110 cm^{-1}.

7. The higher than usual C-H stretching peak at 2940 cm^{-1} is due to the presence of the oxygen.

CONFIRMATION

CAS Index Name: Propane, 2,2'-oxybis-
Molecular Formula: $C_6H_{14}O$
Line Formula: $(CH_3)_2CHOCH(CH_3)_2$
CAS Registry Number: 108-20-3
Beilstein Reference: 1^4, 1471
Molecular Weight: 102.17

Solubility: H_2O 2; al 5; eth 5; ace 3; ctc 3
Melting Point [°C]: -86.8
Boiling Point [°C]: 68.5
Refractive Index: 1.3679^{20}
Density [g/cm^3]: 0.7241^{20}

9

Sample Compound: Acetic Acid

GIVEN: Major peaks from five spectrometric instruments (spectral data only).
FIND: Identity of Compound C

MS Reference: **NIST 34542**
MS Peaks (Intensities): 43(100) 45(87) 60(57) 15(42) 42(14) 29(13) 14(13)
   28(7) 18(6) 16(6)
m/e for Dominant Isotopic Species: 60.02
IR Reference: **COB 4819**
IR Peaks [cm^{-1}]: 3000 2930 2650 1710 1410 1350 1290 940 620
UV Reference: **OES 4-3**
UV Peaks [nm] (Absorp. Coef.): 208(32) EtOH
^{13}C NMR Reference: **JJ 7**
^{13}C NMR Shifts [ppm]: FT 20.6 178.1 CDCl$_3$
^{1}H NMR Reference: **VAR 8**
^{1}H NMR Shifts [ppm]: 2.1 11.4 CDCl$_3$

SOLUTION

This problem illustrates how the problem-solving algorithm may be used when
only spectral data (not graphical spectra) are available.

1. Identify the molecular weight by choosing the largest m/e ratio of some
   intensity from the series of MS peaks. The m/e = 60 is the molecular ion in
   this instance.

2. The even molecular weight indicates a molecular formula having zero or an
   even number of nitrogen atoms.

3. Since M+2 data are absent; no remarkable heteroatoms need to be subtracted
   before using the "Molecular Formula Compilation Table", a condensed
   version of Beynon's Table.

4. The ^{1}H NMR shifts show two different hydrogen atoms. The ^{13}C NMR shifts
   indicate at least two different carbon atoms.

5. The likely candidates include: $C_2H_4O_2$, $C_2H_8N_2$, and C_3H_8O.

6. The respective LU calculations follow:
   $$LU = 2+1-4/2=1, LU = 2+1-8/2+2/2=0, LU = 3+1-8/2=0.$$

7. The molecular formula having the two nitrogens can be eliminated because a compound with two nitrogens having a level of unsaturation equaling zero and two carbons could only be a double primary amine, which would have an IR absorption in the 3500—3400 cm^{-1} range. The highest frequency for this unknown compound is 3000 cm^{-1}.

8. The molecular formula having the single O can also be eliminated because the alcohol would also have an absorption that would be a higher frequency than this unknown has. The possible ether with an odd number of carbons would not have symmetry and so the ^{13}C NMR would have three shifts not two as in the present case.

9. The strongest prospect is $C_2H_4O_2$ having one level of unsaturation. The IR absorptions of 3000 cm^{-1} and 1710 cm^{-1} are good indicators of a carboxylic acid. The base peak of the MS corresponds to the fragment ion $CH_3C=O$. The unknown compound is acetic acid. The second largest peak of the MS at m/e = 45 corresponds to COOH. The ^{1}H NMR with two hydrogens is also consistent with the structure of acetic acid. The proton attached to the OH portion of the carboxylic acid has a chemical shift in the 11.4 range.

CONFIRMATION
CAS Index Name: Acetic acid
Molecular Formula: $C_2H_4O_2$
Line Formula: CH_3CO_2H
CAS Registry Number: 64-19-7
Beilstein Reference: 2^4, 94
Molecular Weight: 60.05
Melting Point [°C]: 16.6
Boiling Point [°C]: 117.9
Density [g/cm^3]: 1.0492^{20}
Refractive Index: 1.3720^{20}
Solubility: H_2O 5; al 5; eth 5; ace 5; bz 5; chl 3; os 3; CS_2 3

Spectroscopic Aids

Spectroscopic Methods

MS Spectrum

The mass spectrum is a plot in which the x-axis represents the mass to charge ratio, and the y-axis represents the relative number of ions. A primary use of this spectrum is the determination of the molecular weight which is numerically equivalent to the mass to charge ratio of the molecular ion.

IR Spectrum

The infrared spectrophotometer records the percent transmittance of incident light through the sample as a function of its wavelength. The presence or absence of major functional groups is detected because the absorption band of a functional group is characteristic of that group regardless of the other structural features of the molecule. Its position lies within an identifying range. Subtleties in placement are due to electronic effects, resonance, and hydrogen bonding.

^{1}H NMR and ^{13}C NMR Spectrum

The number, position, intensity, and splitting pattern of signals in the nuclear magnetic spectrum provide information about the symmetry, electronic environment of the proton or carbon atom, quantity of protons present, and nature of adjacent protons in the molecule. The graph plots the chemical shift in ppm versus the intensity as a percentage of the base peak.

UV Spectrum

The ultraviolet spectrum describes the position of the maximum absorptions on the x-axis and the intensity of the absorbance of the incident light on the y-axis. The $n \rightarrow \pi^*$ and $\pi \rightarrow \pi^*$ transitions are the most observed and useful transitions in organic molecules. The spectrum reveals the degree of unsaturation and quantitatively measures the extent of conjugation of the compound.

MS Spectrum

Gases:

- Transfer the measured amount of the gas sample from a gas bulb into the molecular leak and subsequently to the ionizing chamber.

Liquids:

- Introduce the liquid sample by touching a micropipette to a sintered glass disc or an orifice under gallium or mercury.

- Alternately, the liquid sample may be injected with a hypodermic needle through a serum cap.

Solids and Less Volatile Liquids:

- Sample size ranges from less than a microgram to several milligrams.

- A heated inlet system is used for these compounds.

IR Spectrum

Gases and Low Boiling Liquids:

- The spectra of gases of low boiling liquids may be obtained by expansion of the sample into an evacuated cell.

- Heating the cell increases the effectiveness.

- A rapid scan IR spectrophotometer determines the IR as the gas emerges from a gas chromatograph.

Liquids:

- Liquids may be examined neat or in solution.

- Neat liquids are examined between salt plates without a spacer.

Solids:

- Solids are examined as a mull or a pressed disc, which are both deposited on glassy film.

^{1}H NMR Spectrum

- Using a 60-MHz proton spectrophotometer, use 5—50 mg of the sample in about 0.4 ml of a solvent which ideally contains no proton, has a low boiling point, and is non-polar and inert. A sample holder positions the sample in proper alignment with respect to the main magnetic field, the transmitter coil, and the receiver coil.

UV Spectrum

Vapor phase:

- Quartz cells with path lengths from 1—100 mm are equipped with gas inlets. Cell jackets may be necessary for temperature control.

Solutions:

- Quartz cells, 1-cm^2, are used. A 3-ml solution is required. Clean cells are essential and the cell should be seasoned with the appropriate solvent before the sample solution is introduced.

Units in Spectroscopy

MS Spectrum

Because MS is a plot of m/e versus percentage of base peaks intensities, this method is dimensionless.

IR Spectrum

Wave Number

        1 Kayser = 1 K

        1 kilokayser = 1 kK

        1 reciprocal centimeter = $1cm^{-1}$

        $1 cm^{-1} = 1 K$

        $1 kK = 1000 cm^{-1}$

Wavelength

        1 micron = 1 μ

        1 meter = 1 m = 10^6 μ

        1 m = 10^6 μ

        1 m = 100 cm

^{1}H NMR and ^{13}C NMR Spectrum

Chemical Shift (δ)

        parts per million = ppm

        cycles per second = cps

        1 reciprocal second = $1 s^{-1}$

        1 Hertz = 1 Hz = 1 cps

        1 megahertz = 1 MHz

        $1 cps = 1 s^{-1}$

UV Spectrum

Wavelength

        1 nanometer = 1 nm

        1 m = 10^9 nm

        1 Angstrom = 1 Å

        10^9 nm = 10^{10} Å

Concentration

        molarity = M = moles per liter

Cell length

        1 m = 100 cm

Useful Equations

<u>MS Spectrum</u>

$$R = M/\Delta M$$

R = Resolution
M = Higher mass number of 2 peaks
ΔM = Difference between 2 mass numbers

$$\% (M+1) = 100[(M+1)]/[M]$$

% (M+1) = Calculated value of M+1 when only C, H, N, O, F, P, and I
                 are present
(M+1) = Molecular ion + 1
M = Molecular ion

$$\% (M+2) = 100[(M+2)]/[M]$$

% (M+2) = Calculated value of M+1 when only C, H, N, O, F, P, and I
                 are present
(M+2) = Molecular ion + 2
M = Molecular ion

<u>IR Spectrum</u>

Application of Hooke's Law to approximate IR stretching frequencies:

$$\upsilon = f^{\frac{1}{2}}/(2 \pi c)[\ M_x + M_y/\ M_xM_y]^{\frac{1}{2}}$$

υ = Vibrational frequency in cm^{-1}
c = Speed of light constant, 3×10^8 m/s = 3×10^{10} cm/s
f = Force constant of bond in dynes/cm
M_x = Mass of atom x in grams
M_y = Mass of atom y in grams

$$\text{Wave number} = 1/\lambda\ (10{,}000)$$

λ = Wavelength in microns

19

^{1}H NMR and ^{13}C NMR Spectrum

$$\tau = 10.00 - \delta$$

τ = Tau scale in NMR
δ = Chemical shift measured in ppm

$$\delta = \Delta\upsilon \times 10^6/\text{oscillator frequency (cps)}$$

δ = Chemical shift measured in ppm
$\Delta\upsilon$ = Difference in absorption frequencies of the sample and the reference in cps

$$N = (Na/Nt)\,(Ht)$$

N = Number of protons in molecular formula due to that chemical shift
Na = Number of squares marked off by the electronic integrator in the ^{1}H NMR due to proton/protons, type a
Nt = Total number of squares marked off by the electronic integrator for all chemical shifts
Ht = Total number of protons in the molecular formula

UV Spectrum

$$A = \log Io/I$$

A = Absorbance
Io = Intensity of incident light
I = Intensity of Beer/Lambert Law

$$\varepsilon = A/Cd$$

ε = Molal extension coefficient
C = concentration in molarity
d = path length in centimeters

Predominant Features in the Spectra

(What strikes you when you first eyeball the spectra)

Resist the temptation to analyze every peak in the spectrum, concentrate on major points first.

<u>MS Spectrum</u>

1. Notice whether the molecular ion is odd or even. Apply the Nitrogen Rule.

2. Closely inspect the M+2 peak; a large isotope contribution to M+2 indicates the presence of remarkable heteroatoms such as sulfur (4.4% of molecular ion), chlorine (1/3 as large as molecular ion), or bromine (equal in intensity to the molecular ion).

3. The pattern of the MS is indicative of an alkyl or aromatic compound. Alkyl compounds have a spectrum that resembles a normal curve with a right hand tail. C_3 and C_4 fragments predominate. The aromatic compound produces a right-hand skewed curve with the largest intensity peaks occurring after m/e = 77.

<u>IR Spectrum</u>

Look for the presence or absence of a few major functional groups, particularly C=O, OH, NH, and NH_2.

1. The carbonyl is one of the strongest peaks in the spectrum. It's difficult to miss it. This peak appears in the general region from 1820—1660 cm^{-1}.

> Carboxylic Acid:
> If there is also a broad absorption near 3400 cm^{-1} indicating an OH is also present, think carboxylic acid.

> Amide:
> If there is also a medium absorption near 3500 cm^{-1} indicating an NH is present or perhaps a double peak with equivalent halves indicating a NH_2, think amide.

> Ester:
> If there is also a strong intensity absorption near 1300—1000 cm^{-1} indicating a C-O is present, think ester.

Anhydride:
If there are two C=O absorptions near 1810 and 1760 cm⁻¹, think
anhydride.

Aldehydes and Ketones:
If the carbonyl band is not in tandem with any of the aforementioned
functional groups, think aldehyde or ketone.

2. Hydrogen bonding makes the OH absorption very distinctive. This
absorption represents the one of the broadest of bands.

Alcohols and Phenols:
Check for a broad absorption near 3600—3300 cm⁻¹. Also notice the
peak for the C-O near 1300—1000 cm⁻¹.

Alcohols and phenols are easily distinguished in the low range of the
spectrum. Aromatic bands predominate in the 900—700 cm⁻¹ range.

3. Sharp absorptions characterize triple bonds.

Nitriles and Alkynes:
For nitriles, C≡N, a medium, sharp absorption appears near 2250 cm⁻¹.
For alkynes, C≡C, a weak, sharp absorption is present near 2150 cm⁻¹.

¹H NMR and ¹³C NMR Spectrum

1. Count the number of chemical shifts. The number of ¹³C chemical shifts
indicates the number of different carbon atoms in the molecule. The number
of chemical shifts in the proton spectra tells the number of different protons

2. Is there a shift at the far left of the spectrum? Think aldehyde.

3. Is there a proton shift around 7 ppm or a ¹³C shift around 125 ppm? An
aromatic compound is indicated.

4. Are there distinctive splitting patterns in the ¹H NMR such as the one an
isopropyl group would produce?

- The intensity of the molecular ion is often an indication of the nature of the unidentified compound. More intense peaks allow for easier recognition of the molecular or parent ion.

- The importance of the intensities of the M+1 and M+2 peaks relative to the molecular ion in identifying remarkable heteroatoms is also of great significance.

Easy to detect

- Aromatics
- Conjugated alkenes
- Saturated ring compounds
- Some sulfur compounds
- Short chain hydrocarbons

Moderate detection

- Straight chain ketones
- Straight chain esters
- Straight chain carboxylic acids
- Straight chain aldehydes
- Straight chain amides
- Straight chain ethers
- Straight chain halides

Difficult to detect

- Aliphatic alcohols
- Aliphatic amines
- Aliphatic nitrates
- Aliphatic nitrites
- Nitroalkanes
- Alkyl nitriles
- Highly branched hydrocarbons

Intensities of Remarkable Heteroatom Isotope Peaks
(in Relation to the Molecular Ion)

Heteroatom	%M+1	%M+2	%M+4
Br	0	98	—
2 Br	0	195	96
3 Br	0	293	286
Br, Cl	130	32	—
Br, 2 Cl	163	74	—
2 Br, Cl	228	158	—
Cl	0	33	—
2 Cl	0	65	11
3 Cl	0	98	32
F	0	—	—
I	0	—	—
P	0	—	—
S	1	4	—

Molecular Formula Compilation Table

This table is utilized after heteroatoms other than oxygen and nitrogen have been subtracted from the given molecular ion. Atoms which have been subtracted are referred to as remarkable heteroatoms in this *Guide*.

Candidate Molecular Formula	M+1 Peak	M+2 Peak
Molecular Ion = 41		
C_3H_5	3.32	0.04
Molecular Ion = 44		
C_2H_4O	2.26	0.21
C_2H_6N	2.64	0.02
C_3H_8	3.37	0.04
Molecular Ion = 45		
C_2H_5O	2.28	0.21
C_2H_7N	2.65	0.02
Molecular Ion = 46		
CH_4NO	1.57	0.21
C_2H_6O	2.30	0.22
Molecular Ion = 47		
CH_3O_2	1.21	0.41
Molecular Ion = 48		
CH_4O_2	1.22	0.40
Molecular Ion = 51		
C_4H_3	4.37	0.07
Molecular Ion = 52		
C_3H_2N	3.66	0.05
C_4H_4	4.39	0.07
Molecular Ion = 53		
C_3H_3N	3.67	0.05
C_4H_5	4.40	0.07

Candidate Molecular Formula	M+1 Peak	M+2 Peak
Molecular Ion = 54		
C_3H_2O	3.31	0.24
C_3H_4N	3.69	0.05
C_4H_6	4.42	0.07
Molecular Ion = 55		
C_3H_3O	3.33	0.24
C_3H_5N	3.70	0.05
C_4H_7	4.43	0.07
Molecular Ion = 56		
C_3H_4O	3.34	0.24
C_3H_6N	3.72	0.05
C_4H_8	4.45	0.08
Molecular Ion = 57		
C_2H_3NO	2.63	0.22
C_3H_5O	3.36	0.24
C_3H_7N	3.74	0.05
C_4H_9	4.47	0.08
Molecular Ion = 58		
$C_2H_2O_2$	2.27	0.42
$C_2H_6N_2$	3.02	0.03
C_3H_6O	3.38	0.24
C_4H_{10}	4.48	0.08
Molecular Ion = 59		
$C_2H_3O_2$	2.29	0.42
C_2H_5NO	2.66	0.22
C_3H_7O	3.39	0.24
C_3H_9N	3.77	0.05
Molecular Ion = 60		
$C_2H_8N_2$	3.05	0.03
C_3H_8O	3.41	0.24

Molecular Formula Compilation Table (Continued)

Candidate Molecular Formula	M+1 Peak	M+2 Peak
Molecular Ion = 65		
C_4H_3N	4.75	0.09
C_5H_5	5.48	0.12
Molecular Ion = 66		
C_4H_4N	4.77	0.09
C_5H_6	5.50	0.12
Molecular Ion = 67		
C_4H_3O	4.41	0.27
C_4H_5N	4.78	0.09
C_5H_7	5.52	0.12
Molecular Ion = 68		
C_4H_4O	4.43	0.28
C_5H_8	5.53	0.12
Molecular Ion = 69		
C_4H_5O	4.44	0.28
C_4H_7N	4.82	0.09
C_5H_9	5.55	0.12
Molecular Ion = 70		
C_4H_6O	4.46	0.28
C_4H_8N	4.83	0.09
C_5H_{10}	5.56	0.13
Molecular Ion = 71		
C_4H_7O	4.47	0.28
Molecular Ion = 72		
C_4H_8O	4.49	0.28
C_5H_{12}	5.60	0.13

27

Candidate Molecular Formula	M+1 Peak	M+2 Peak
Molecular Ion = 73		
$C_2H_3NO_2$	2.67	0.42
C_3H_7NO	3.77	0.25
C_4H_9O	4.51	0.28
$C_4H_{11}N$	4.88	0.09
Molecular Ion = 74		
$C_4H_{10}O$	4.52	0.28
Molecular Ion = 75		
$C_3H_7O_2$	3.43	0.44
C_3H_9NO	3.81	0.25
C_6H_3	6.53	0.18
Molecular Ion = 76		
$C_3H_8O_2$	3.45	0.44
Molecular Ion = 77		
C_5H_3N	5.83	0.14
C_6H_5	6.56	0.18
Molecular Ion = 78		
C_6H_6	6.58	0.18
Molecular Ion = 79		
C_5H_3O	5.49	0.32
C_5H_5N	5.86	0.14
C_6H_7	6.60	0.18
Molecular Ion = 80		
C_6H_8	6.61	0.18
Molecular Ion = 81		
C_5H_7N	5.90	0.14
C_6H_9	6.63	0.18

Candidate Molecular Formula	M+1 Peak	M+2 Peak
Molecular Ion = 82		
C_5H_6O	5.54	0.32
C_6H_{10}	6.64	0.19
Molecular Ion = 83		
C_5H_9N	5.93	0.15
C_6H_{11}	6.66	0.19
Molecular Ion = 84		
C_5H_8O	5.57	0.33
$C_5H_{10}N$	5.94	0.15
C_6H_{12}	6.68	0.19
Molecular Ion = 85		
$C_5H_{11}N$	5.96	0.15
C_6H_{13}	6.69	0.19
Molecular Ion = 86		
$C_3H_6N_2O$	4.14	0.27
$C_5H_{12}N$	5.98	0.15
$C_4H_6O_2$	4.50	0.48
C_6H_{14}	6.71	0.19
Molecular Ion = 87		
$C_5H_{13}N$	5.99	0.15
C_6HN	6.88	0.20
Molecular Ion = 100		
C_8H_4	8.71	0.33
Molecular Ion = 101		
C_7H_3N	7.99	0.28
C_8H_5	8.72	0.33

Candidate Molecular Formula	M+1 Peak	M+2 Peak
Molecular Ion = 102		
$C_3H_6N_2O_2$	4.18	0.47
$C_4H_6O_3$	4.54	0.68
$C_5H_{10}O_2$	5.64	0.53
$C_6H_{14}O$	6.75	0.39
C_8H_6	8.74	0.34
Molecular Ion = 103		
C_7H_5N	8.03	0.28
C_8H_7	8.76	0.34
Molecular Ion = 104		
C_7H_4O	7.67	0.45
C_7H_6N	8.04	0.28
C_8H_8	8.77	0.34
Molecular Ion = 105		
C_8H_9	8.79	0.34
Molecular Ion = 108		
C_7H_8O	7.73	0.46
C_8H_{12}	8.84	0.34
Molecular Ion = 109		
$C_7H_{11}N$	8.12	0.29
C_8H_{13}	8.85	0.35
Molecular Ion = 110		
$C_7H_{10}O$	7.76	0.46
$C_7H_{12}N$	8.14	0.29
C_8H_{14}	8.87	0.35
Molecular Ion = 111		
$C_7H_{11}O$	7.78	0.46
C_8H_{15}	8.88	0.35
Molecular Ion = 112		
$C_7H_{12}O$	7.80	0.46
C_8H_{16}	8.90	0.35

Molecular Formula Compilation Table (Continued)

Candidate Molecular Formula	M+1 Peak	M+2 Peak
Molecular Ion = 113		
C_9H_5	9.81	0.43
Molecular Ion = 114		
$C_7H_{14}O$	7.83	0.47
C_8H_{18}	8.93	0.35
Molecular Ion = 115		
C_8H_5N	9.11	0.37
C_9H_7	9.84	0.43
Molecular Ion = 116		
C_9H_6	9.85	0.43
Molecular Ion = 117		
C_9H_9	9.87	0.43
Molecular Ion = 118		
$C_5H_{10}O_3$	5.68	0.73
$C_6H_{14}O_2$	6.79	0.60
C_8H_6O	8.78	0.54
C_9H_{10}	9.89	0.44
Molecular Ion = 119		
C_7H_5NO	8.06	0.48
C_8H_7O	8.80	0.54
C_8H_9N	9.17	0.37
C_9H_{11}	9.90	0.44
Molecular Ion = 120		
C_8H_8O	8.81	0.54
C_9H_{12}	9.92	0.44
Molecular Ion = 121		
C_9H_{13}	9.93	0.44
Molecular Ion = 122		
$C_8H_{10}O$	8.84	0.54
C_9H_{14}	9.95	0.44

Candidate Molecular Formula	M+1 Peak	M+2 Peak
Molecular Ion = 134		
$C_7H_2O_3$	7.71	0.86
$C_7H_6N_2O$	8.46	0.52
$C_8H_6O_2$	8.82	0.74
$C_8H_{10}N_2$	9.57	0.41
$C_9H_{10}O$	9.92	0.64
$C_{10}H_{14}$	11.03	0.55
Molecular Ion = 135		
C_8H_9NO	9.21	0.58
$C_{10}H_{15}$	11.05	0.55
Molecular Ion = 136		
$C_{10}H_{16}$	11.06	0.55
Molecular Ion = 137		
$C_7H_7NO_2$	8.14	0.69
$C_8H_{11}NO$	9.24	0.58
$C_{10}H_{17}$	11.08	0.56
Molecular Ion = 138		
$C_8H_{10}O_2$	8.88	0.75
$C_9H_{14}O$	9.99	0.60
$C_{10}H_{18}$	11.09	0.56
Molecular Ion = 143		
$C_8H_{17}NO$	9.34	0.59
$C_{19}H_{19}O$	10.07	0.65
$C_9H_{21}N$	10.44	0.49
$C_{10}H_9N$	11.33	0.58
$C_{11}H_{11}$	12.06	0.66
Molecular Ion = 144		
$C_{10}H_8O$	10.97	0.74
$C_{11}H_{12}$	12.08	0.67
$C_6H_{14}N_3O$	7.89	0.47
$C_8H_2NO_2$	9.14	0.77
$C_{10}H_8O$	10.97	0.74
$C_{11}H_{12}$	12.08	0.67

Molecular Formula Compilation Table (Continued)

Candidate Molecular Formula	M+1 Peak	M+2 Peak
Molecular Ion = 145		
$C_5H_7NO_4$	6.05	0.96
$C_6H_9O_4$	6.78	1.00
$C_7H_{13}O_3$	7.89	0.87
$C_8H_{19}NO$	9.37	0.59
Molecular Ion = 146		
$C_5H_8NO_4$	6.07	0.96
$C_6H_{12}NO_3$	7.17	0.82
$C_7H_2N_2O_2$	8.44	0.71
$C_7H_{14}O_3$	7.91	0.87
Molecular Ion = 147		
$C_4H_{11}N_4O_2$	6.10	0.56
$C_6H_3N_4O$	8.10	0.49
$C_6H_{11}O_4$	6.82	1.00
$C_7H_{15}O_3$	7.92	0.87
Molecular Ion = 148		
$C_5H_{10}NO_4$	6.10	0.96
$C_5H_{16}N_4O$	7.22	0.43
$C_6H_{12}O_4$	6.83	1.00
$C_7H_8N_4$	9.22	0.38
Molecular Ion = 149		
$C_4H_{13}N_4O_2$	6.13	0.56
$C_6H_3N_3O_2$	7.75	0.66
$C_9H_9O_2$	9.95	0.84
$C_{11}H_3N$	12.32	0.69
Molecular Ion = 150		
$C_5H_{12}NO_4$	6.13	0.96
$C_7H_8N_3O$	8.88	0.55
$C_9H_{12}NO$	10.34	0.68
$C_{11}H_2O$	11.96	0.85

Candidate Molecular Formula	M+1 Peak	M+2 Peak
Molecular Ion = 151		
$C_5H_3N_4O_2$	7.06	0.62
$C_7H_5NO_3$	8.14	0.89
$C_8H_{11}N_2O$	9.62	0.62
$C_{10}HNO$	11.24	0.77
Molecular Ion = 152		
$C_6H_4N_2O_3$	7.43	0.84
$C_7H_{10}N_3O$	8.91	0.55
$C_9H_2N_3$	10.90	0.54
$C_{10}H_2NO$	11.26	0.78
Molecular Ion = 153		
$C_5H_5N_4O_2$	7.09	0.62
$C_7H_5O_4$	7.80	1.07
$C_8H_9O_4$	8.91	0.95
$C_{11}H_7N$	12.38	0.70
Molecular Ion = 154		
$C_6H_4NO_4$	7.09	1.02
$C_8H_2N_4$	10.20	0.47
$C_9H_{18}N_2$	10.78	0.53
$C_{11}H_6O$	12.02	0.86
Molecular Ion = 155		
$C_5H_5N_3O_3$	6.75	0.80
$C_7H_7O_4$	7.83	1.07
$C_7H_{15}N_4$	9.33	0.39
$C_{11}H_{23}$	12.26	0.69
Molecular Ion = 156		
$C_7H_8O_4$	7.85	1.07
$C_8H_4N_4$	10.24	0.47
$C_8H_{16}N_2O$	9.70	0.62
$C_{12}H_{12}$	13.16	0.80

Molecular Formula Compilation Table (Continued)

Candidate Molecular Formula	M+1 Peak	M+2 Peak
Molecular Ion = 157		
$C_8H_{13}N_4O$	8.26	0.50
$C_7H_9O_4$	7.87	1.07
$C_8H_{19}N_3$	10.09	0.46
$C_9H_{19}NO$	10.45	0.69
Molecular Ion = 158		
$C_5H_{10}N_4O_2$	7.17	0.63
$C_7H_{10}O_4$	7.88	1.07
$C_8H_6N_4$	10.27	0.48
$C_9H_{18}O_2$	10.09	0.86
Molecular Ion = 159		
$C_6H_9NO_4$	7.17	1.02
$C_7H_{17}N_3O$	9.02	0.56
$C_8H_{19}N_2O$	9.75	0.63
$C_{10}H_7O_2$	11.00	0.95
Molecular Ion = 160		
$C_5H_8N_2O_4$	6.45	0.98
$C_8H_8N_4$	10.30	0.48
$C_9H_4O_3$	9.91	1.04
$C_{12}H_{16}$	13.22	0.80
Molecular Ion = 161		
$C_6H_{11}NO_4$	7.20	1.03
$C_7H_3N_3O_2$	8.84	0.75
C_8HO_4	8.82	1.14
$C_9H_{11}N_3$	11.05	0.56
Molecular Ion = 162		
$C_6H_{14}N_2O_3$	7.59	0.85
$C_8H_2O_4$	8.83	1.15
$C_{10}H_{10}O_2$	11.04	0.95
$C_{12}H_2O$	13.04	0.98

Candidate Molecular Formula	M+1 Peak	M+2 Peak
Molecular Ion = 163		
$C_6HN_3O_3$	7.76	0.87
$C_7H_{15}O_4$	7.96	1.08
$C_{11}H_{17}N$	12.54	0.72
$C_{12}H_{19}$	13.27	0.81
Molecular Ion = 164		
$C_6H_2N_3O_3$	7.78	0.87
$C_7H_{16}O_4$	7.98	1.08
$C_9H_{10}NO_2$	10.35	0.88
$C_{11}H_4N_2$	12.71	0.74
Molecular Ion = 165		
$C_6H_3N_3O_3$	7.79	0.87
$C_8H_5O_4$	8.88	1.15
$C_{10}H_3N_3O$	12.00	0.66
$C_{11}H_{17}O$	12.20	0.88
Molecular Ion = 166		
$C_6H_4N_3O_3$	7.81	0.87

Levels of Unsaturation (LU)

Determination of the level of unsaturation provides an excellent insight into the structure of the unknown compound. The equation used for this calculation is stated below:

LU = number of carbon atoms + 1 − number of hydrogen atoms/2 − number of halogen atoms/2 + trivalent nitrogen atoms/2

Number of Unsaturation Sites	Possible Combinations
LU = 1	1 Double bond
	1 Ring
LU = 2	1 Triple bond
	1 Ring, 1 double bond
	2 Rings
	2 Double bonds
LU = 3	3 Double bonds
	1 Ring + 2 double bonds
LU = 4	1 Aromatic ring

Diagnostic Fragment Ions in the Mass Spectrum

Mass:Charge Ratio	Ion
26	$C\equiv N$
29	C_2H_5 CHO
30	CH_2NH_2
31	CH_2OH OCH_3
41	$CH_2C\equiv N$
43	C_3H_7 $CH_3C=O$
45	$C(=O)OH$
47	CH_2SH
57	C_4H_9
59	$C(=O)OCH_3$
60	$CH_2C(=O)OH$
61	CH_2CH_2SH CH_2SCH_3
71	C_5H_{11}
74	$CH_2C(=O)OCH_3$
77	C_6H_5
83	C_4H_3S
85	C_6H_{13}
91	$C_6H_5CH_2$
94	C_6H_5O
105	$C_6H_5C=O$

Diagnostic IR Absorptions Frequencies

Functional Group	Approximate Frequency (cm^{-1})
O-H for alcohols	3400
O-H for carboxylic acids	3100
N-H	3400
C-H for alkynes	3300
C-H for alkenes	3050
C-H for alkanes	2900
C=C	1650
C=O for amides	1690
C=O for carboxylic acids	1715
C=O for aldehyde and ketones	1730
C=O for esters	1740
C=O for acid halides	1790
C=O for acid anhydrides	1820 and 1750
C≡C	2150
C≡N	2260
Benzene with 1 substituent	750 and 700
Benzene with ortho disubstitution	750
Benzene with para disubstitution	815
Benzene with meta disubstitution	780 and 705
C-O-C	1120
C(=O)-O	1200

Diagnostic Infrared Absorptions

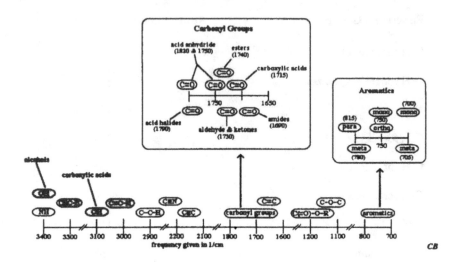

40

Wavelength to Wave Number Conversion Table

Wavelength (μm)	Wave number (cm^{-1})									
	0	1	2	3	4	5	6	7	8	9
2.0	5000	4975	4950	4926	4902	4878	4854	4831	4808	4785
2.1	4762	4739	4717	4695	4673	4651	4630	4608	4587	4566
2.2	4545	4525	4505	4484	4464	4444	4425	4405	4386	4367
2.3	4348	4329	4310	4292	4274	4255	4237	4219	4202	4184
2.4	4167	4149	4232	4115	4098	4082	4065	4049	4032	4016
2.5	4000	3984	3968	4953	3937	3922	3006	3891	3876	3861
2.6	3846	3831	3817	3802	3788	3774	3759	3745	3731	3717
2.7	3704	3690	3676	3663	3650	3636	3623	3610	3597	3584
2.8	3571	3559	3546	3534	3521	3509	3497	3484	3472	3460
2.9	3448	3436	3425	3413	3401	3390	3378	3367	3356	3344
3.0	3333	3322	3311	3300	3289	3279	3268	3257	3247	3236
3.1	3226	3215	3205	3195	3185	3175	3165	3155	3145	3135
3.2	3125	3115	3106	3096	3086	3077	3067	3058	3049	3040
3.3	3030	3021	3012	3003	2994	2985	2976	2967	2959	2950
3.4	2941	2933	2924	2915	2907	2899	2890	2882	2874	2865
3.5	2857	2849	2841	2833	2825	2817	2809	2801	2793	2786
3.6	2778	2770	2762	2755	2747	2740	2732	2725	2717	2710
3.7	2703	2695	2688	2681	2674	2667	2660	2653	2646	2639
3.8	2632	2625	2618	2611	2604	2597	2591	2584	2577	2571
3.9	2654	2558	2551	2545	2538	2532	2525	2519	2513	2506
4.0	2500	2494	2488	2481	2475	2469	2463	2457	2451	2445
4.1	2439	2433	2427	2421	2415	2410	2404	2398	2387	2387
4.2	2381	2375	2370	2364	2358	2353	2347	2342	2336	2331
4.3	2326	2320	2315	2309	2304	2299	2294	2288	2283	2278
4.4	2273	2268	2262	2257	2252	2247	2242	2237	2232	2227
4.5	2222	2217	2212	2208	2203	2198	2193	2188	2183	2179
4.6	2174	2169	2165	2160	2155	2151	2146	2141	2137	2132
4.7	2128	2123	2119	2114	2110	2105	2101	2096	2092	2088
4.8	2083	2079	2075	2070	2066	2062	2058	2053	2049	2045
4.9	2041	2037	2033	2028	2024	2020	2016	2012	2008	2004
5.0	2000	1996	1992	1988	1984	1980	1976	1972	1969	1965
5.1	1961	1957	1953	1949	1946	1942	1938	1934	1931	1927
5.2	1923	1919	1916	1912	1908	1905	1901	1898	1894	1890
5.3	1887	1883	1880	1876	1873	1869	1866	1862	1859	1855
5.4	1852	1848	1845	1842	1838	1835	1832	1828	1825	1821
5.5	1818	1815	1812	1808	1805	1802	1799	1795	1792	1788
5.6	1786	1783	1779	1776	1773	1770	1767	1764	1761	1757
5.7	1754	1751	1748	1745	1742	1739	1736	1733	1730	1727
5.8	1724	1721	1718	1715	1712	1709	1706	1704	1701	1698
5.9	1695	1692	1689	1686	1684	1681	1678	1675	1672	1669

Wavelength (μm)	Wave number (cm⁻¹)									
	0	1	2	3	4	5	6	7	8	9
6.0	1667	1664	1661	1568	1656	1653	1650	1647	1645	1642
6.1	1639	1637	1634	1631	1629	1626	1623	1621	1618	1616
6.2	1613	1610	1608	1605	1603	1600	1597	1595	1592	1590
6.3	1587	1585	1582	1580	1577	1575	1572	1570	1567	1565
6.4	1563	1560	1558	1555	1553	1550	1548	1546	1543	1541
6.5	1538	1536	1534	1531	1529	1527	1524	1522	1520	1517
6.6	1515	1513	1511	1508	1506	1504	1502	1499	1497	1495
6.7	1493	1490	1488	1486	1484	1481	1479	1477	1475	1473
6.8	1471	1468	1466	1464	1462	1460	1458	1456	1453	1451
6.9	1449	1447	1445	1443	1441	1439	1437	1435	1433	1431
7.0	1429	1427	1425	1422	1420	1418	1416	1414	1412	1410
7.1	1408	1406	1404	1403	1401	1399	1397	1395	1393	1391
7.2	1389	1387	1385	1383	1381	1379	1377	1376	1374	1372
7.3	1370	1368	1366	1364	1362	1361	1359	1357	1355	1353
7.4	1351	1350	1348	1346	1344	1342	1340	1339	1337	1335
7.5	1333	1332	1330	1328	1326	1325	1323	1321	1319	1318
7.6	1316	1314	1312	1311	1309	1307	1305	1304	1302	1300
7.7	1299	1297	1295	1294	1292	1290	1289	1287	1285	1284
7.8	1282	1280	1279	1277	1276	1274	1272	1271	1269	1267
7.9	1266	1264	1263	1261	1259	1258	1256	1255	1253	1252
8.0	1250	1248	1247	1245	1244	1242	1241	1239	1238	1236
8.1	1235	1233	1232	1230	1229	1227	1225	1224	1222	1221
8.2	1220	1218	1217	1215	1214	1212	1211	1209	1208	1206
8.3	1205	1203	1202	1200	1199	1198	1196	1195	1193	1192
8.4	1190	1189	1188	1186	1185	1183	1182	1181	1179	1178
8.5	1176	1175	1174	1172	1171	1170	1168	1167	1166	1164
8.6	1163	1161	1160	1159	1157	1156	1155	1153	1152	1151
8.7	1149	1148	1147	1145	1144	1143	1142	1140	1139	1138
8.8	1136	1135	1134	1133	1131	1130	1129	1127	1126	1125
8.9	1124	1122	1121	1120	1119	1117	1116	1115	1114	1112
9.0	1111	1110	1109	1107	1106	1105	1104	1103	1101	1100
9.1	1099	1098	1096	1095	1094	1093	1092	1091	1089	1088
9.2	1087	1086	1085	1083	1082	1081	1080	1079	1078	1076
9.3	1075	1074	1073	1072	1071	1070	1068	1067	1066	1065
9.4	1064	1063	1062	1060	1059	1058	1057	1056	1055	1054
9.5	1053	1052	1050	1049	1048	1047	1046	1045	1044	1043
9.6	1042	1041	1040	1038	1037	1036	1035	1034	1033	1032
9.7	1031	1030	1029	1028	1027	1026	1025	1024	1022	1021
9.8	1020	1019	1018	1017	1016	1015	1014	1013	1012	1011
9.9	1010	1009	1008	1007	1006	1005	1004	1003	1002	1001

Wavelength to Wave Number Conversion Table (Continued)

Wavelength (μm)	Wave number (cm⁻¹)									
	0	1	2	3	4	5	6	7	8	9
10.0	1000	999	998	997	996	995	994	993	992	991
10.1	990	989	988	987	986	985	984	983	982	981
10.2	980	979	978	978	977	976	975	974	973	972
10.3	971	970	969	968	967	966	965	964	963	962
10.4	962	961	960	959	958	957	956	955	954	953
10.5	952	951	951	950	949	948	947	946	945	944
10.6	943	943	942	041	940	939	938	937	936	935
10.7	935	934	933	932	931	930	929	929	928	927
10.8	926	925	924	923	923	922	921	920	919	918
10.9	917	917	916	915	914	913	912	912	911	910
11.0	909	908	907	907	906	905	904	903	903	902
11.1	901	900	899	898	898	897	896	895	894	894
11.2	893	892	891	890	890	889	888	887	887	886
11.3	885	884	883	883	882	881	880	880	879	878
11.4	877	876	876	875	874	873	873	872	871	870
11.5	870	869	868	867	867	866	865	864	864	863
11.6	862	861	861	860	859	858	858	857	856	855
11.7	855	854	853	853	852	851	850	850	849	848
11.8	847	847	846	845	845	844	843	842	842	841
11.9	840	840	839	838	838	837	836	835	835	834
12.0	833	833	832	831	831	830	829	829	828	827
12.1	826	826	825	824	824	823	822	822	821	820
12.2	820	819	818	818	817	816	816	815	814	814
12.3	813	812	812	811	810	810	809	808	808	807
12.4	806	806	805	805	804	803	803	802	801	801
12.5	800	799	799	798	797	797	796	796	795	794
12.6	794	793	792	792	791	791	790	789	789	788
12.7	787	787	786	786	785	784	784	783	782	782
12.8	781	781	780	779	779	778	778	777	776	776
12.9	775	775	774	773	773	772	772	771	770	770
13.0	769	769	768	767	767	766	766	765	765	764
13.1	763	763	762	762	761	760	760	759	759	758
13.2	758	757	756	756	755	755	754	754	753	752
13.3	752	751	751	750	750	749	749	748	747	747
13.4	746	746	745	745	744	743	743	742	742	741
13.5	741	740	740	739	739	738	737	737	736	736
13.6	735	735	734	734	733	733	732	732	731	730
13.7	730	729	729	728	728	727	727	726	726	725
13.8	725	724	724	723	723	722	722	721	720	720
13.9	719	719	718	718	717	717	716	716	715	715

^{1}H NMR

1. The different kinds of hydrogen atoms is indicated.

2. Integration gives the number of protons. Peak areas are proportional to that number; a direct ratio is the rule.

3. The splitting pattern is defined as (n+1) where n = number of protons on neighboring carbon atoms. The relative intensities of the peaks of a multiplet depend upon n. A singlet is 1; a doublet is 1:1; a triplet is 1:2:1; a quartet is 1:3:3:1.

4. The general region of chemical shifts is 0—11 ppm. The position of the chemical shift depends on the proton being inspected, not the one/ones that cause the splitting pattern.

5. Electronegative elements do influence the chemical shift. Protons attached to carbons absorb at less than 2 ppm; a proton attached to a carbon that also is attached to a nitrogen atom absorbs at 2.5, and a proton attached to a carbon that also is attached to an oxygen atom absorbs at 3.5 ppm.

^{13}C NMR

1. The number of different kinds of carbon atoms is indicated.

2. A large intensity peak indicates more carbon atoms are present, but there is no direct ratio.

3. The (n+1) rule refers to the number of protons attached directly to the carbon being inspected. A quaternary carbon, one with no attached protons, produces a singlet. A tertiary carbon, one with a single proton attached, gives a doublet. A secondary carbon, one with two protons, yields a triplet. A primary carbon having three hydrogens attached gives a quartet.

4. The general region of chemical shifts 0—200 ppm. The position of the chemical shift depends on the carbon being inspected.

5. Electronegative elements influence shifts to a lesser degree due to the larger relative general region of chemical shifts.

Approximate Chemical Shifts in ^{1}H NMR

Type of Proton	Chemical Shift (ppm)
Protons attached to primary carbons	1
Protons attached to secondary carbons	1.3
Protons attached to tertiary carbons	1.6
Alpha protons attached to the carbonyl carbon Allylic protons	2
Benzylic protons Protons attached to a carbon bonded to a nitrogen	2.5
Protons attached to an sp carbon as in alkynes	3
Protons attached to a carbon bonded to an oxygen Protons attached to a carbon bonded to a halogen	3.5
Vinylic protons	5—6
Aromatic protons	7—8
Aldehydic protons	9—10
Proton component of carboxylic acid	11

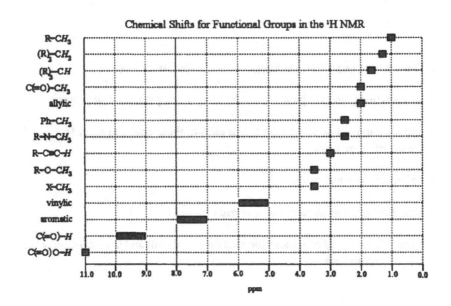

Chemical Shifts for Functional Groups in the ^{1}H NMR

Integration of ^{1}H NMR

The peak areas are proportional to the number of protons they represent. Graphically, the area is equal to the number of squares. Alternately, the area can be determined by using the length in mm. Proton counting with an electronic tracer is often superimposed as it is in the ^{1}H NMR of 1,2,3-trichloropropane.

The analyst focuses on major signals and averages out complex multiplets. Please refer to the figure below.

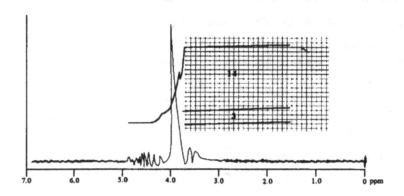

1. Draw a base line where the integrator has begun its tracing.

2. At each peak, find the point where the integrator's path has a slope near zero.

3. Draw a parallel line to the base line at each of these points in order to simplify the counting process.

4. Label the chemical shifts.

5. Count the protons corresponding to each.

6. Record the number of squares for each chemical shift and the total number of squares for all peaks.

7. Write an expression for each chemical shift that represents the relationship between the peak area ratios and the H's in the molecular formula.

$$N = (Na/Nt) (Ht)$$

    N = Number of protons in molecular formula due to that chemical shift
    Na = Number of squares marked off by the electronic integrator in the
           ^{1}H NMR due to proton/protons, type a
    Nt = Total number of squares marked off by the electronic integrator
           for all chemical shifts
    Ht = Total number of protons in the molecular formula

8. The total number of squares is 17; 14 are due to the 4 terminal protons, and 3 are due to the single proton on the secondary carbon.

3/17 x 5 protons = .88 ≈ 1 proton   14/17 x 5 protons = 4.1 ≈ 4 protons

9. The intensity (height) of irregularly shaped doublet is due to four protons; the intensity of the multiplet is due to one proton.

Approximate Chemical Shifts in ^{13}C NMR

Type of Carbon	Chemical Shift (ppm)
Primary carbon	0—35
Secondary carbon	15—40
Primary carbon with a bromine attached	20—40
Tertiary carbon	25—50
Primary carbon with chlorine attached	25—50
Primary carbon with amino group attached	35—50
Primary carbon with hydroxyl group attached	50—65
Alkyne carbon	65—90
Alkene carbon	100—150
Aromatic carbon	110—175
Carbonyl carbon	190—220

UV Absorptions of Chromophores

Chromophore	Typical Compound	λ_{max}, mμ nm[a]	ε_{max}[b]
C=C	Ethylene	171	15,530
C≡C	2-Octyne	178	10,000
		196	ca 2,100
		223	160
RC(=O)H	Acetaldehyde	160	20,000
		180	10,000
		290	17
RC(=O)R'	Acetone	166	16,000
		189	900
		279	15
RC(=O)OH	Acetic acid	208	32
RC(=O)Cl	Acetyl chloride	220	100
RC(=O)NH$_2$	Acetamide	178	9,500
		220	63
RC(=O)OR'	Ethyl acetate	211	57
RNO$_2$	Nitromethane	201	5,000
		274	17
C=C-C=C	Butadiene	217	20,900
C=C-C(=O)H	Crotonaldehyde	218	18,000
C(NO$_2$)=C	1-Nitro-1-propene	229	9,400

[a] Wavelength of maximum absorption.
[b] Molar extinction coefficient.

Glossary

Mass Spectrum

Base Peak: Most intense peak.

Beynon's Table: Molecular formula compilation corresponding to masses.

Level of Unsaturation: Number of unsaturation sites equals $C+1-H/2-X/2+N/2$.

Mass Spectrum: Graphical representation of the masses of the positively charged fragments versus their relative concentrations.

Molecular Ion: Ion produced when a molecule loses an electron.

Nitrogen Rule: A molecule of even numbered molecular weight must contain zero or even numbers of nitrogen atoms.

Relative Intensity: Intensity of the peak relative to the base peak, which is assigned an intensity of 100.

Unsaturation Site: A double bond or ring has one unsaturation site; a triple bond has two unsaturation sites, and a benzene ring has four unsaturation sites.

Infrared Spectrum

Fingerprint Region: Pattern of peaks in the 1300—625 cm^{-1} region which is specific to the compound.

Infrared Spectrum: Plot of the IR radiation transmitted as a function of the wavelength or wavenumber.

Wave Number: IR units in reciprocal centimeters that are proportional to frequency.

Ultraviolet Spectrum

B Band: Benzenoid bands that are broad absorption bands between 230 and 270 nm.

Blue Shift: Hypsochromic shift describing a shift to a shorter wavelength due to substitution or solvent effect.

Chromophore: Functional group that is responsible for $\pi \rightarrow \pi^*$ and or $n \rightarrow \pi^*$ transitions.

R Band: $n \rightarrow \pi^*$ transition.

Nuclear Magnetic Spectrum

Broadband Decoupling: Technique in ^{13}C NMR used to simplify interpretation in which all the proton-carbon couplings are removed.

^{13}C NMR: Spectroscopy which utilizes the ^{13}C isotope to reflect the environments of individual carbon atoms in a molecule.

Chemical Shifts: Shifts in the position of NMR absorptions that arise from shielding and deshielding by electrons.

Coupling Constant: Constant which quantitatively describes the extent that two nuclear spins are coupled; constant that often equals the distance between adjacent peaks in a split NMR signal.

Downfield: Descriptive term referring to the low field region of the spectrum which lies to the left of the spectrum.

Integrated Area: Area of a signal in a ^{1}H NMR spectrum which is proportional to the number of equivalent protons responsible for the peak.

Multiplicity: Number of peaks present in the NMR due to the splitting of the signal, which follows the n+1 rule where n represents the number of adjacent protons.

Shielding: Effect of the electrons in a molecules which causes a decrease in the strength of an external magnetic field felt by a proton.

Representative Compounds

GIVEN: MS spectrum, m/e for dominant isotopic species, IR spectrum, ¹H NMR chemical shifts, UV absorptions
FIND: Identity of Compound # 1

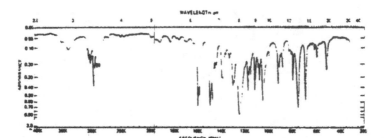

Figure 5. IR compound #1. (With permission from Bio-Rad/Sadtler Division.)

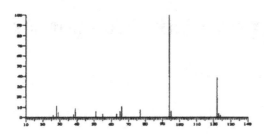

Figure 6. MS compound #1.

MS Reference: **NIST 62358**
MS Peaks (Intensities): 94(100) 122(39) 28(12) 66(11) 39(9) 77(8) 95(7) 65(7) 51(7) 29(6)
m/e for Dominant Isotopic Species: 122.07
IR Reference: **COB 6191**
IR Peaks [cm⁻¹]: 3040 3000 2940 1600 1580 1500 1480 1390 1300 1240 1170 1120 1050 920 880 800 750 690
UV Reference: **SAD 77**
UV Peaks [nm] (Absorp. Coef.): 278(1190) 271(1400) 220(6400) MeOH
¹H NMR Reference: **SAD 26**
¹H NMR Shifts [ppm]: 1.3 3.9 6.9 CCl₄

SOLUTION

This problem is a good one for the beginning analyst.

1. The ^{1}H NMR at 6.9 ppm clearly indicates an aromatic compound. The MS spectrum affirms this idea with the presence of the prominent molecular ion. The base peak, m/e = 94, typifies the $C_6H_6O^+$ fragment. IR peaks at 1240 cm^{-1} and 1050 cm^{-1} strengthen the hypothesis that the unknown is an aralkyl ether.

2. Candidates selected from the "Molecular Formula Compilation Table" should have greater than six carbons, have significant unsaturation, contain at least one oxygen atom, and have zero or an even number of nitrogen atoms. Only two such molecular formulas fit the profile postulated: $C_7H_6O_2$ (LU = 7+1-6/2=5) and $C_8H_{10}O$ (LU = 8+1-10/2=4).

3. The ^{1}H NMR pattern favors the latter formula in that there are at least two different protons that are not aromatic. The aralkyl ether notion is further substantiated by the chemical shift at 3.9 ppm, the absorption that characterizes a proton attached to a carbon atom bonded to an oxygen atom.

4. When the phenoxy ion is subtracted from the stipulated molecular formula, an ethyl group remains ($C_8H_{10}O-C_6H_5O=C_2H_5$).

5. Inspection of the ^{1}H NMR shows the triplet and quartet pattern, verifying the ethyl group.

CONFIRMATION

CAS Index Name: Benzene, ethoxy-
Molecular Formula: $C_8H_{10}O$
Line Formula: $C_6H_5OC_2H_5$
CAS Registry Number: 103-73-1
Beilstein Reference: 6^4, 554
Molecular Weight: 122.16

Solubility: H_2O 1; al 3; eth 3; ctc 3
Melting Point [°C]: -29.5
Boiling Point [°C]: 169.8
Refractive Index: 1.5076^{20}
Density [g/cm^3]: 0.9666^{20}

GIVEN: IR spectrum, MS spectrum, m/e, ^{13}C NMR chemical shifts
FIND: Identity of Compound #2

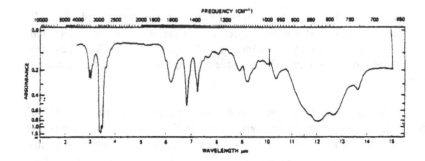

Figure 7. IR compound #2. (With permission from
Bio-Rad/Sadtler Division.)

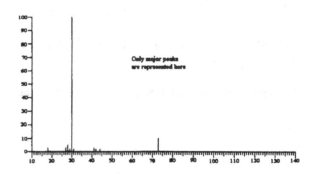

Only major peaks
are represented here

Figure 8. MS compound #2.

MS Reference: **NIST 51519 WILEY 581**
MS Peaks (Intensities): 30(100) 73(10) 28(5) 41(3) 27(3) 18(3) 44(2) 42(2)
   31(2) 29(2)
m/e for Dominant Isotopic Species: 73.09
IR Reference: **SADG 8485**
IR Peaks [cm^{-1}]: 3380 3300 2960 2930 2850 1610 1460 1370 1090 980
  840 790
^{13}C NMR Reference: **STOT 152**
^{13}C NMR Shifts [ppm]: 14.0 20.4 36.7 42.3

SOLUTION

1. The IR spectrum features a weak bifurcated absorption in the 3300 cm^{-1} range; a primary amine is suspected. A major peak at m/e of 30 further points to a $CH_2NH_2^+$ ion. Four different kinds of carbon atoms are indicated by the ^{13}C NMR chemical shifts.

2. The molecular weight of the unknown is 73. The odd molecular weight indicates that the compound has an uneven number of nitrogen atoms in the compound.

3. Inspection of the M+1 and M+2 indicates the absence atoms other than those addressed in Beynon's Table.

4. The only possible molecular formula candidate from the "Molecular Formula Compilation Table" is $C_4H_{11}N$.

5. The level of unsaturation is calculated. LU = 4+1-11/2+1/2=0.

6. The only structure that would give an ^{13}C NMR indicating four different carbon atoms is the straight chain butyl amine $CH_3CH_2CH_2CH_2NH_2$.

CONFIRMATION
CAS Index Name: 1-Butanamine
Molecular Formula: $C_4H_{11}N$
Line Formula: $CH_3CH_2CH_2CH_2NH_2$
CAS Registry Number: 109-73-9
Beilstein Reference: 4⁴, 540
Molecular Weight: 73.13
Melting Point [°C]: -49.1
Boiling Point [°C]: 77.0
Density [g/cm³]: 0.7414[20]
Refractive Index: 1.4031[20]
Solubility: H_2O 5; al 3; eth 3

GIVEN: IR spectrum, MS spectrum, molecular weight equals 118. This molecular weight is given because the molecular ion is too weak to be read from the MS.

FIND: Identity of Compound #3

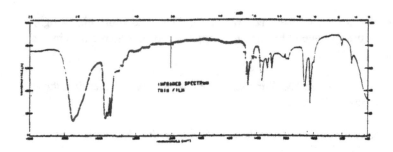

Figure 9. IR compound #3. (With permission from Bio-Rad/Sadtler Division.)

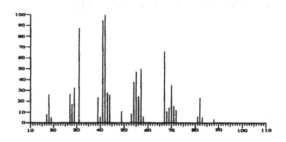

Figure 10. MS compound #3.

MS Reference: **NIST 63584**

MS Peaks (Intensities): **42(100) 41(95) 31(87) 67(66) 57(50) 55(47) 54(37) 70(35) 29(33) 43(28)**

m/e for Dominant Isotopic Species: 118.10

IR Reference: **SADG 21349**

IR Peaks [cm⁻¹]: 3350 2920 2860 1460 1430 1050 720 650

¹H NMR Reference: **SILVERSTEIN 424**

¹H NMR Shifts [ppm]: 1.4 2.0 3.6

SOLUTION

1. A weak molecular ion may be indicative of an alcohol and points to a low level of unsaturation. The strong peak at m/e equal to 31 reinforces the alcohol suspicion. Further validation is provided by the strong and broad absorption at 3400 cm^{-1} in the IR spectrum. ^{1}H NMR shows three different protons.

2. Zero or even number of nitrogen atoms are required as per the even molecular weight.

3. No heteroatoms other than nitrogen and oxygen need be considered.

4. Characteristic nitrogen absorption is absent from the IR.

5. The likely molecular formula candidates are: $C_4H_6O_4$, $C_5H_{10}O_3$, and $C_6H_{14}O_2$.

6. LU = 4+1-6/2=2, 5+1-10/2=1, and 6+1-14/2=0, respectively.

7. IR inspection eliminates carbonyl absorption. Absence of absorptions at around 1650 cm^{-1} and 2100 cm^{-1} preclude double and triple bonds.

8. The most promising molecular formula is the saturated one. Although an important M-18 peak is missing, the M-36 peak at m/e = 36 is present.

9. Only 3 different protons shifts in a compound having 14 hydrogens have a great deal of symmetry.

10. The MS spectrum has a straight chain pattern.

CONFIRMATION

CAS Index Name: 1,6-Hexanediol
Molecular Formula: $C_6H_{14}O_2$
Line Formula: HO(CH$_2$)$_6$OH
CAS Registry Number: 629-11-8
Beilstein Reference: 1^4, 2556
Molecular Weight: 118.17

Solubility: H$_2$O 3; al 3; eth 2; ace 3; bz 1
Melting Point [°C]: 42.8
Boiling Point [°C]: 208
Refractive Index: 1.4579^{25}

GIVEN: IR spectrum, MS spectrum, ^{1}H NMR chemical shifts
FIND: Identity of Compound #4

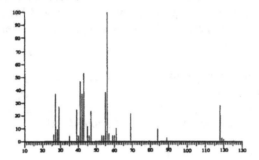

Figure 11. MS compound #4.

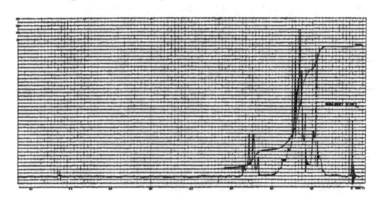

Figure 12. ^{1}H NMR compound #4. (With permission from Bio-Rad/Sadtler Division.)

MS Reference: **NIST 2067 WILEY 463**
MS Peaks (Intensities): 56(100) 43(54) 41(47) 55(38) 42(37) 27(37) 118(28) 29(27) 47(24) 69(22)
m/e for Dominant Isotopic Species: 118.08
IR Reference: μ **SADP 6364**
IR Peaks [cm^{-1}]: 2940 2560 1470 1370 1280 1270 1220 760 730
UV Reference: **OES 2-79**
UV Peaks [nm] (Absorp. Coef.): 224(126) cyhex
^{1}H NMR Reference: **SAD 18712**
^{1}H NMR Shifts [ppm]: 0.9 1.2 1.5 2.5 CCl$_4$

SOLUTION

1. Notice the significant M+2 peak in the MS, e.g. 4.96 intensity conjures up the presence of sulfur due to the 4.4% abundance of the ^{34}S isotope. Cleavage of the C-C bond adjacent to the S-H bond is the rule; the CH_2SH ion resulting in a characteristic peak at m/e equal to 47. The base peak at 57 represents the butyl ion. Aliphatic mercaptans show S-H stretching absorption in the 2600—2550 cm^{-1} range. In the ^1H NMR sulfhydryl proton absorbs in the 1.2—1.6 ppm range.

2. The molecular weight is 118. There are three different protons according to the proton NMR. Zero or even number of nitrogen atoms according to the Nitrogen Rule.

3. The unknown represents the presence of a remarkable heteroatom, namely sulfur. The weight is corrected before choosing candidates, 118-32=86. Likely possibilities from the "Molecular Formula Compilation Table" include: $C_2H_6N_4$, $C_3H_6N_2O$, $C_4H_6O_2$, $C_4H_{10}N_2$, $C_5H_{10}O$, C_6H_{14}.

4. Inspection of the IR absorptions show an absence of carbonyl (approximately 1700 cm^{-1}), alkene (approximately 1699 cm^{-1}) and alkyne (approximately 2200 cm^{-1}).

5. Level of unsaturation should be a good way to eliminate several candidates.

   LU = 2+1-6/2+4/2=2;   LU = 3+1-6/2+2/2=2
   LU = 4+1-6/2=2;   LU = 4+1-10/2+2/2=1
   LU = 5+1-10/2=1;   LU = 6+1-14/2=0

6. The straight chain thiol fits the aforementioned conditions as well as the typical straight chain pattern of the MS with peaks at m/e equal to 29, 43, 57, and 71.

CONFIRMATION

CAS Index Name: 1-Hexanethiol
Molecular Formula: $C_6H_{14}S$
Line Formula: $CH_3(CH_2)_4CH_2SH$
CAS Registry Number: 111-31-9
Beilstein Reference: 1^4, 1705
Molecular Weight: 118.23

Solubility: H_2O 1; al 4; eth 4
Melting Point [°C]: -81
Boiling Point [°C]: 151
Refractive Index: 1.4496^{20}
Density [g/cm^3]: 0.8424^{20}

GIVEN: IR spectrum, MS spectrum, ^{1}H NMR and ^{13}C chemical shifts.
FIND: Identity of Compound #5

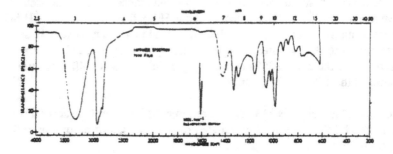

Figure 13. IR compound #5. (With permission from
Bio-Rad/Sadtler Division.)

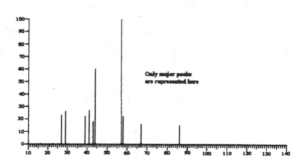

Figure 14. MS compound #5.

MS Reference: **NIST 57753**; **WILEY 557**
MS Peaks (Intensities): 57(100) 44(60) 41(27) 29(26) 27(23) 58(22) 39(22)
  43(18) 67(16) 86(15)
m/e for Dominant Isotopic Species: 86.07
IR Reference: **SADG 12974**
IR Peaks [cm^{-1}]: 3350 2950 2860 1440 1340 1300 1260 1070 1030 990
  830 640
^{13}C NMR Reference: **STOT 176**
^{13}C NMR Shifts [ppm]: 23.7 35.3 73.6
^{1}H NMR Reference: **SAD 44**
^{1}H NMR Shifts [ppm]: 1.6 3.6 4.2 CCl$_4$

62

SOLUTION

1. Notice the IR absorption at 3350 cm^{-1} indicating the presence of an OH or NH. The IR spectrum does not suggest unsaturation. The ^{13}C chemical shifts show a minimum of three carbon atoms.

2. The molecular ion has a mass to charge ratio equal to 86. This compound has zero or an even number of nitrogen atoms.

3. No special heteroatoms are apparent from M+1 and M+2 peak intensities.

4. Being cognizant of the aforementioned parameters, the possible candidates are: $C_3H_6N_2O$; $C_4H_6O_2$; $C_5H_{10}O$.

5. The LU's are calculated respectively:
   $$LU = 3+1-6/2+2/2=2; LU = 4+1-6/2=2; LU = 5+1-10/2=1.$$

6. The most likely candidate is the one with least unsaturation. $C_5H_{10}O-OH=C_5H_9$. Because the C=C bond absorption was absent in the IR, perhaps, the unsaturation site is due to a ring.

7. The ^{1}H NMR chemical shifts are in the alkyl and OH range.

8. The base peak of the MS spectrum, m/e = 57, fits the proposed scenario. The $C_3H_5O^+$ is typical of a cyclic alcohol, resulting from a ring cleavage mechanism.

9. Although there are five carbon atoms in the molecule, there are only three ^{13}C chemical shifts. The molecule exhibits significant symmetry.

CONFIRMATION

CAS Index Name: Cyclopentanol
Molecular Formula: $C_5H_{10}O$
CAS Registry Number: 96-41-3
Beilstein Reference: 6^4, 5
Molecular Weight: 86.13
Refractive Index: 1.4530^{20}

Solubility: H_2O 2; al 3; eth 3; ace 3; ctc 2
Melting Point [°C]: -19
Boiling Point [°C]: 140.4
Density [g/cm^3]: 0.9478^{20}

GIVEN: IR spectrum, MS spectrum, ^{1}H NMR chemical shifts.
FIND: Identity of Compound #6

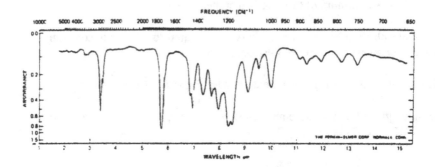

Figure 15. IR compound #6. (With permission from
Bio-Rad/Sadtler Division.)

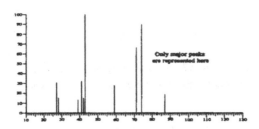

Figure 16. MS compound #6.

MS Reference: **NIST 62383 WILEY 270**
MS Peaks (Intensities): 43(100) 74(90) 71(66) 41(32) 27(31) 59(28) 87(19)
    42(15) 28(15) 39(14)
m/e for Dominant Isotopic Species: 102.07
IR Reference: μ **SADP 337**
IR Peaks [cm^{-1}]: 2940 2860 1750 1450 1370 1320 1270 1190 1100 1010
    880 790 750
^{1}H NMR Reference: **SAD 6669**
^{1}H NMR Shifts [ppm]: 0.9 1.6 2.2 3.6 CCl$_4$

SOLUTION

1. An ester is a strong prospect in viewing the IR spectrum. The typical strong absorption at 1750 cm^{-1} is a carbonyl band at a higher frequency than a normal ketone. The absorption at 1190 cm^{-1} is characteristic of the ester C(=O)O functional group.

2. The molecular weight is 102, indicating zero or an even number of nitrogen atoms. Remarkable heteroatoms are absent.

3. The ^{1}H NMR spectrum indicates four different hydrogen atoms.

4. Using the "Molecular Formula Compilation Table", the likely candidates have at least two oxygen atoms. Their respective levels of unsaturation are:

$C_3H_6N_2O_2$	LU = 3+1-6/2+2/2=2
$C_5H_{10}O_2$	LU = 5+1-10/2=1
$C_4H_6O_3$	LU = 4+1-6/2=2

5. The molecular formula with one level of unsaturation seems likely; additional unsaturation sites are precluded by the absence of absorptions at 1650 cm^{-1} (C=C) or 2200 cm^{-1} (C≡C).

6. Another strong indicator of the ester, in particular a methyl ester, is found in the MS spectrum. The m/e peak equal to 59 is typical of an aliphatic, unbranched methyl ester which has resulted from a McLafferty rearrangement and cleavage one bond removed from the carbonyl group. The 43 base peak is the other fragment resulting from this cleavage. The RC≡O$^+$, m/e = 31, is also a diagnostic peak for methyl esters.

7. When the methyl ester functional group is subtracted, $(C_5H_{10}O_2-C_2H_3O_2=C_3H_7)$, the remaining carbon chain is necessarily a straight chain evidenced by three ^{1}H NMR chemical shifts.

CONFIRMATION

CAS Index Name: Butanoic acid, methyl ester
Molecular Formula: $C_5H_{10}O_2$
Line Formula: $CH_3CH_2CH_2CO_2CH_3$
CAS Registry Number: 623-42-7
Beilstein Reference: 2^4, 786

Solubility: H$_2$O 2; al 5; eth 5; ctc 2
Melting Point [°C]: -85.8
Boiling Point [°C]: 102.8
Molecular Weight: 102.13
Density [g/cm^3]: 0.8984[20]
Refractive Index: 1.3878[20]

GIVEN: MS peaks and respective intensities, chemical shifts for NMR, other spectral data.

FIND: Identity of Compound #7

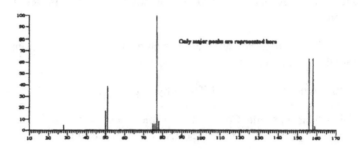

Figure 17. MS compound #7.

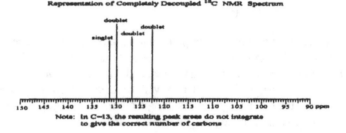

Figure 18. ^{13}C NMR compound #7.

MS Reference: **NIST 61413**

MS Peaks (Intensities): 77(100) 158(64) 156(64) 51(39) 50(17) 78(8) 76(6) 75(6) 28(5) 159(4)

m/e for Dominant Isotopic Species: 155.96

IR Reference: μ **COB 6476**

IR Peaks [cm⁻¹]: 3030 1590 1470 1450 1080 1020 1000 910 740 690

UV Reference: **SAD 954**

UV Peaks [nm] (Absorp. Coef.): 271(1190) 264(1710) 261(1730) 257(1590) 250(1390) MeOH

^{13}C NMR Reference: **JJ 152**

^{13}C NMR Shifts [ppm]: FT 122.4 126.7 129.8 131.4 CDCl$_3$

^{1}H NMR Reference: **SAD 1212**

^{1}H NMR Shifts [ppm]: 7.1 7.4 CCl$_4$

SOLUTION

This compound illustrates the procedure followed when remarkable heteroatoms are present. When elements other than carbon, hydrogen, oxygen, or nitrogen are present, their exact number must be determined so that their masses can be subtracted from the molecular weight before using a Beynon-type table.

1. Notice the molecular ion and the M+2 ion are of equal intensity. Peaks m/e equal to 158 and 156 are both 64% of the base peak. A clear cut indication of the presence of bromine results from equivalent natural abundances of ^{79}Br and ^{81}Br.

2. The molecular weight is 156, the even molecular weight implying a molecule having zero or an even number of nitrogen atoms.

3. The proton NMR shows two chemical shifts, both aromatic protons. The ^{13}C NMR spectrum points to four distinguishable carbon atoms.

4. Before using the collection of molecular formulas, subtract the atomic mass of 79 from the 156 (molecular ion peak). The remainder equals 77; the only molecular formula that satisfies the aformentioned conditions is C_6H_5.

CONFIRMATION
CAS Index Name: Benzene, bromo-
Molecular Formula: C_6H_5Br
CAS Registry Number: 108-86-1
Beilstein Reference: 5[4], 670
Molecular Weight: 156.00
Melting Point [°C]: -30.6
Boiling Point [°C]: 156.0
Density [g/cm³]: 1.4950[20]
Refractive Index: 1.5597[20]
Solubility: H_2O 1; al 4; eth 4; bz 4; ctc 3

GIVEN: IR spectrum, UV absorptions, NMR chemical shifts.
FIND: Identity of Compound #8

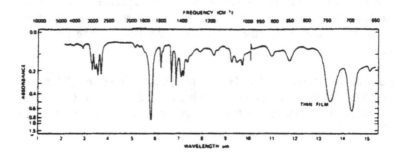

Figure 19. IR compound #8. (With permission from
Bio-Rad/Sadtler Division.)

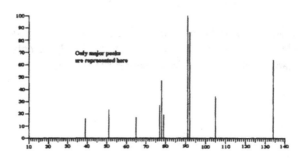

Figure 20. MS compound #8.

MS Reference: **NIST 20930**
MS Peaks (Intensities): 91(100) 92(86) 134(64) 78(47) 105(34) 77(27) 51(23)
   79(19) 65(17) 39(16)
m/e for Dominant Isotopic Species: 134.07
IR Reference: μ **SADP 17410**
IR Peaks [cm⁻¹]: 3030 2940 1720 1610 1490 1450 1410 1140 1040 740 690
UV Reference: **SAD 5535**
UV Peaks [nm] (Absorp. Coef.): 268 264 261 258 253 247 242 208 MeOH
^{13}C NMR Reference: **JJ 343**
^{13}C NMR Shifts [ppm]: FT 28.0 45.0 126.1 128.2 128.4 140.4 201.1 CDCl$_3$
^{1}H NMR Reference: **VAR 529**
^{1}H NMR Shifts [ppm]: 2.7 3.0 7.2 9.8 CDCl$_3$

68

SOLUTION

1. Strong indicators of the carbonyl functional group are present in the IR (1720 cm^{-1}) and ^{1}H NMR (9.8 ppm) spectra.

2. MS gives a molecular weight of 134 with a very distinctive peak of 91, a benzyl prospect. The abundance of UV absorptions substantiates this idea.

3. The candidates with appropriate unsaturation, having a minimum of one oxygen atom, zero or even numbers of nitrogen, and at least seven carbon atoms indicated by the seven chemical shifts in the ^{13}C NMR are:

$C_7H_2O_3$	LU = 7+1-2/2=7
$C_7H_6N_2O$	LU = 7+1-6/2+2/2=6
$C_8H_6O_2$	LU = 8+1-6/2=6
$C_9H_{10}O$	LU = 9+1-10/2=5

4. A benzene ring and a carbonyl group account for five sites of unsaturation. The last possibility fits these parameters. $C_9H_{10}O\text{-CHO}=C_8H_9\text{-}C_6H_5=C_2H_4$.

5. Ten protons in the molecular formula yield only four chemical shifts in the ^{1}H NMR.

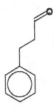

CONFIRMATION
CAS Index Name: Benzenepropanal
Molecular Formula: $C_9H_{10}O$
Line Formula: $C_6H_5CH_2CH_2CHO$
CAS Registry Number: 104-53-0
Beilstein Reference: 7[4], 692
Molecular Weight: 134.17
Melting Point [°C]: 47
Boiling Point [°C]: 223[745], 104-5[13]
Color: mcl
Solubility: H_2O 1; al 4; eth 5

GIVEN: MS spectrum, IR spectrum, NMR chemical shifts, molecular weight = 69.

FIND: Identity of Compound #9

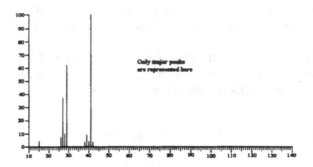

Only major peaks are represented here

Figure 21. MS compound #9.

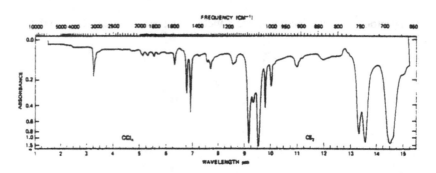

Figure 22. IR compound #9. (With permission from Bio-Rad/Sadtler Division.)

MS Reference: **NIST 46212 WILEY 845**

MS Peaks (Intensities): 41(100) 29(62) 27(37) 28(10) 39(9) 26(7) 40(5) 42(4) 38(4) 15(4)

m/e for Dominant Isotopic Species: 69.06

IR Reference: μ **COB 2557**

IR Peaks [cm⁻¹]: 2940 2860 2220 1470 1430 1390 1330 1100 920 840 740

¹H NMR Reference: **SAD 33**

¹H NMR Shifts [ppm]: 1.1 1.7 2.3 CCl₄

¹³C NMR Reference: **SILVERSTEIN 272**

¹³C NMR Shifts [ppm]: 17.3 100.9 116.0 150.2

SOLUTION

1. The absorption at 2220 cm^{-1} is readily identified as a C≡N. This information coupled with an ^{1}H NMR chemical shift at 2.3 ppm points to a nitrile. The ^{13}C NMR chemical shifts indicate a compound with a least four different carbon atoms.

2. The MS spectrum inspection precludes remarkable heteroatoms.

3. The only candidate with an appropriate molecular formula having a molecular weight of 69 is C$_4$H$_7$N whose level of unsaturation is LU = 4+1-7/2+1/2=2. A nitrile has a level of unsaturation equal to 2.

4. The straight chain nitrile fits the four different carbons condition. Also the base peak of straight chain nitriles between C$_4$ and C$_9$ is m/e equal to 41, resulting from a hydrogen rearrangement. Although this is the case for this unknown compound, its diagnostic value is somewhat limited because the m/e of 41 is also present in many hydrocarbons due to the C$_3$H$_5$ ion.

CONFIRMATION
CAS Index Name: Butanenitrile
Molecular Formula: C$_4$H$_7$N
Line Formula: CH$_3$CH$_2$CH$_2$CN
CAS Registry Number: 109-74-0
Beilstein Reference: 2^4, 806
Molecular Weight: 69.10
Melting Point [°C]: -111.9
Boiling Point [°C]: 117.6
Density [g/cm^3]: 0.7936^{20}
Refractive Index: 1.3842^{20}
Solubility: H$_2$O 2; al 5; eth 5; bz 3; ctc 2

GIVEN: IR Spectrum, MS Spectrum, m/e, NMR chemical shifts
FIND: Identity of Compound #10

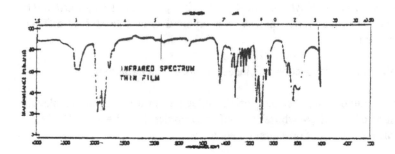

Figure 23. IR compound #10. (With permission from
Bio-Rad/Sadtler Division.)

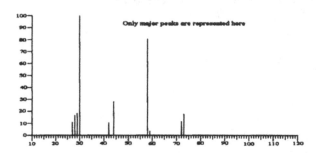

Figure 24. MS compound #10.

MS Reference: **NIST 19154**
MS Peaks (Intensities): 30(100) 58(81) 44(28) 73(18) 29(18) 28(17) 72(12)
    42(11) 27(11) 59(4)
m/e for Dominant Isotopic Species: 73.09
IR Reference: μ **SADP 3194**
IR Peaks [cm^{-1}]: 3330 2940 2860 1450 1390 1320 1140 1040 720
UV Reference: **HBCP**
UV Peaks [nm] (Absorp. Coef.): 222(295) 194(2951) gas
^{13}C NMR Reference: **JJ 98**
^{13}C NMR Shifts [ppm]: FT 15.4 44.1 CDCl$_3$
^{1}H NMR Reference: **SAD 7057**
^{1}H NMR Shifts [ppm]: 0.9 1.0 2.6 CCl$_4$

SOLUTION

1. At first glance, notice the band at 3330 cm^{-1} in the IR, the alkyl pattern in the MS, and three different protons and two different carbon atoms in the NMR spectra.

2. The m/e is 73, indicating the molecular weight allows for a molecular formula with odd numbers of nitrogen atoms.

3. Using the "Molecular Formula Compilation Table" under molecular ion equal to 73, look for a compound with at least two C's, three H's, and one N present.

4. The candidate molecular formulas and their calculated LU's found are $C_2H_3NO_2$ (LU = 2), C_3H_7NO (LU = 1), and $C_4H_{11}N$ (LU = 0).

5. Because indications of unsaturation are absent, the first possibility pursued is the saturated one.

6. The IR band at 3330 cm^{-1}, which is not bifurcated, is typical of a secondary amine. A bifurcated peak is characteristic of primary amines.

7. A secondary amine RNHR' with a molecular formula $C_4H_{11}N$ has two possible configurations, i.e., $C_2H_5NHC_2H_5$ and $C_3H_7NHCH_3$. The two different C's in the ^{13}C NMR point to symmetry in the molecule, thus pointing to the former possibility.

CONFIRMATION
CAS Index Name: Ethanamine, N-ethyl-
Molecular Formula: $C_4H_{11}N$
Line Formula: $(C_2H_5)_2NH$
CAS Registry Number: 109-89-7
Beilstein Reference: 4[4], 313
Molecular Weight: 73.13
Melting Point [°C]: -49.8
Boiling Point [°C]: 55.5
Density [g/cm^3]: 0.7056[20]
Refractive Index: 1.3864[20]
Solubility: H_2O 4; al 5; eth 3; ctc 3

Practice Problems

GIVEN: IR spectrum, MS spectrum, m/e
FIND: Identity of P-1

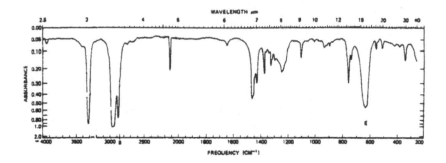

Figure 25. IR P-1. (With permission from Bio-Rad/Sadtler Division.)

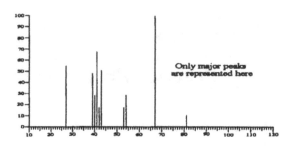

Only major peaks are represented here

Figure 26. MS P-1.

MS Reference: **NIST 455 WILEY 114**
MS Peaks (Intensities): 67(100) 41(67) 27(65) 43(51) 39(48) 54(29) 40(28) 53(17) 42(17) 81(11)
m/e for Dominant Isotopic Species: 82.08
IR Reference: μ **COB 2370**
IR Peaks [cm⁻¹]: 3230 2940 2860 2130 1470 1430 1370 1240 740
UV Reference: **OES 5-74**
UV Peaks [nm] (Absorp. Coef.): 187(447) cyhex
^{13}C NMR Reference: **JJ 175**
^{13}C NMR Shifts [ppm]: FT 13.5 18.1 21.9 30.7 68.1 84.5 CDCl₃
^{1}H NMR Reference: **SAD 8182**
^{1}H NMR Shifts [ppm]: 0.9 1.4 1.7 2.1 CCl₄

76

SOLUTION

Resist the temptation to analyze every peak or chemical shift!

1. What strikes you at first glance when you eyeball spectra and spectral data?

2. Is the molecular weight or m/e odd or even?

3. Are remarkable heteroatoms present?

4. Subtract the weight of remarkable heteroatoms.

5. What are the possible candidates from the "Molecular Formula Compilation Table"?

6. Their calculated LU are:

7. Inspect spectral data and spectra using the diagnostic spectroscopic aids.

8. Subtract the apparent functional groups from the "best prospect molecular formula".

9. Tentative structures are:

10. Reconcile your best candidate with available data.

11. Are there any discrepancies which you can't explain?

GIVEN: IR spectrum, ¹H NMR spectrum, m/e, M+2 = 93% of M, M+4 = 30% of M

FIND: Identity of P-2

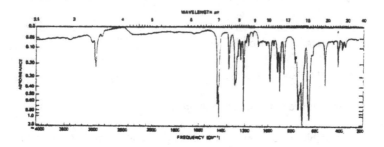

Figure 27. IR P-2. (With permission from Bio-Rad/Sadtler Division.)

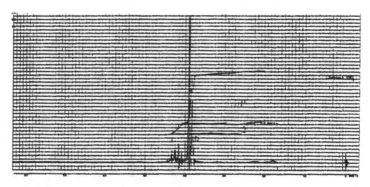

Figure 28. ¹H NMR P-2. (With permission from Bio-Rad/Sadtler Division.)

MS Reference: **NIST 21378**

MS Peaks (Intensities): 75(100) 39(58) 49(42) 110(38) 61(34) 77(33) 112(22) 27(16) 97(15) 38(15)

m/e for Dominant Isotopic Species: 145.95

IR Reference: μ **COB 5859**

IR Peaks [cm⁻¹]: 2940 1450 1330 1280 1250 1220 1180 1150 1100 990 930 910 870 780 750 710 670 660

¹³C NMR Reference: **JJ 23**

¹³C NMR Shifts [ppm]: cw 45.3 59.0 diox

¹H NMR Reference: **SAD 6769**

¹H NMR Shifts [ppm]: 3.9 4.2 CCl₄

SOLUTION

Resist the temptation to analyze every peak or chemical shift!

1. What strikes you at first glance when you eyeball spectra and spectral data?

2. Is the molecular weight or m/e odd or even?

3. Are remarkable heteroatoms present?

4. Subtract the weight of remarkable heteroatoms.

5. What are the possible candidates from the "Molecular Formula Compilation Table"?

6. Their calculated LU are:

7. Inspect spectral data and spectra using the diagnostic spectroscopic aids.

8. Subtract the apparent functional groups from the "best prospect molecular formula".

9. Tentative structures are:

10. Reconcile your best candidate with available data.

11. Are there any discrepancies which you can't explain?

GIVEN: IR spectrum, ^{1}H NMR spectrum, MS peaks, ^{13}C NMR chemical shifts
FIND: Identity of P-3

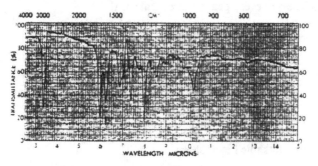

Figure 29. IR P-3. (With permission from
Bio-Rad/Sadtler Division.)

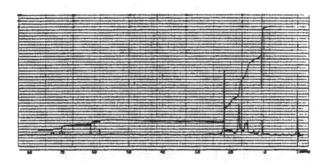

Figure 30. ^{1}H NMR P-3. (With permission
from Bio-Rad/Sadtler Division.)

MS Reference: **NIST 47245 WILEY 1340**
MS Peaks (Intensities): 177(100) 43(53) 93(15) 41(14) 178(13) 123(12)
    91(12) 135(11) 192(10) 121(10)
m/e for Dominant Isotopic Species: 192.15
IR Reference: μ **SADP 5907**
IR Peaks [cm^{-1}]: 2941 1667 1613 1449 1351 1299 1250 1163 1136 1020
    980
UV Reference: **OES 6-501**
UV Peaks [nm] (Absorp. Coef.): 296(10471) 226(6310) EtOH
^{13}C NMR Reference: **STOT 437**
^{13}C NMR Shifts [ppm]: 19.3 21.6 26.9 28.9 33.6 34.3 39.8 132.4 135.0
    136.5 142.0 196.5
^{1}H NMR Reference: **VAR 617**
^{1}H NMR Shifts [ppm]: 1.1 1.8 2.1 2.3 6.1 7.3 CDCl$_3$

SOLUTION

Resist the temptation to analyze every peak or chemical shift!

1. What strikes you at first glance when you eyeball spectra and spectral data?

2. Is the molecular weight or m/e odd or even?

3. Are remarkable heteroatoms present?

4. Subtract the weight of remarkable heteroatoms.

5. What are the possible candidates from the "Molecular Formula Compilation Table"?

6. Their calculated LU are:

7. Inspect spectral data and spectra using the diagnostic spectroscopic aids.

8. Subtract the apparent functional groups from the "best prospect molecular formula".

9. Tentative structures are:

10. Reconcile your best candidate with available data.

11. Are there any discrepancies which you can't explain?

81

GIVEN: MS spectrum, m/e, ^{13}C spectrum, ^{1}H NMR chemical shifts, IR absorptions

FIND: Identity of P-4

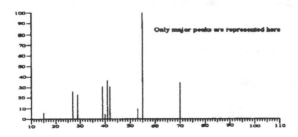

Figure 31. MS P-4.

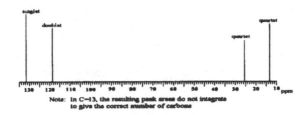

Figure 32. ^{13}C NMR P-4.

MS Reference: **NIST 242 ALDN 162**

MS Peaks (Intensities): 55(100) 41(36) 70(35) 42(31) 39(31) 27(26) 29(24) 53(10) 15(6) 40(5)

m/e for Dominant Isotopic Species: 70.08

IR Reference: **SADG 12975**

IR Peaks [cm^{-1}]: 2980 2940 2880 1450 1370 1220 1160 800

UV Reference: **HBCP**

UV Peaks [nm] (Absorp. Coef.): 205(851) gas

^{13}C NMR Reference: **JJ 119**

^{13}C NMR Shifts [ppm]: cw 13.3 25.5 118.7 131.7 diox

^{1}H NMR Reference: **SAD 3411**

^{1}H NMR Shifts [ppm]: 1.5 1.6 5.1 CCl$_4$

82

SOLUTION

Resist the temptation to analyze every peak or chemical shift!

1. What strikes you at first glance when you eyeball spectra and spectral data?

2. Is the molecular weight or m/e odd or even?

3. Are remarkable heteroatoms present?

4. Subtract the weight of remarkable heteroatoms.

5. What are the possible candidates from the "Molecular Formula Compilation Table"?

6. Their calculated LU are:

7. Inspect spectral data and spectra using the diagnostic spectroscopic aids.

8. Subtract the apparent functional groups from the "best prospect molecular formula".

9. Tentative structures are:

10. Reconcile your best candidate with available data.

11. Are there any discrepancies which you can't explain?

GIVEN: MS spectrum, ^{1}H NMR spectrum, IR absorptions
FIND: Identity of P-5

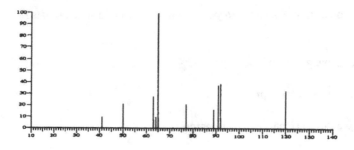

Figure 33. MS P-5.

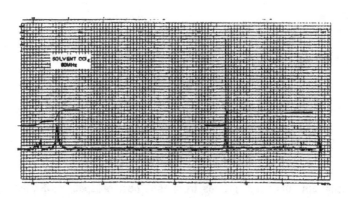

Figure 34. ^{1}H NMR P-5. (With permission
from Bio-Rad/Sadtler Division.)

MS Reference: **NIST 72129 ALDN 86**
MS Peaks (Intensities): 65(100) 92(38) 91(37) 120(34) 63(27) 77(21)
    50(21) 89(16) 64(10) 41(10)
m/e for Dominant Isotopic Species: 137.05
IR Reference: **COB 7426**
IR Peaks [cm^{-1}]: 3085 2970 2930 2850 1615 1580 1525 1485 1465 1435
    1350 1305 1205 1165 1150 1040 950 860 785 725 665
UV Reference: **SAD 1292**
UV Peaks [nm] (Absorp. Coef.): 254(4330) MeOH
^{1}H NMR Reference: **SAD 676**
^{1}H NMR Shifts [ppm]: 2.6 7.3 7.9 CCl$_4$

84

SOLUTION

Resist the temptation to analyze every peak or chemical shift!

1. What strikes you at first glance when you eyeball spectra and spectral data?

2. Is the molecular weight or m/e odd or even?

3. Are remarkable heteroatoms present?

4. Subtract the weight of remarkable heteroatoms.

5. What are the possible candidates from the "Molecular Formula Compilation Table"?

6. Their calculated LU are:

7. Inspect spectral data and spectra using the diagnostic spectroscopic aids.

8. Subtract the apparent functional groups from the "best prospect molecular formula".

9. Tentative structures are:

10. Reconcile your best candidate with available data.

11. Are there any discrepancies which you can't explain?

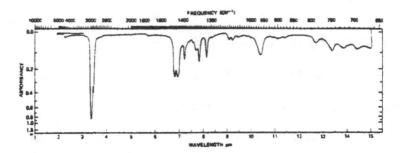

Figure 35. IR P-6. (With permission from
Bio-Rad/Sadtler Division.)

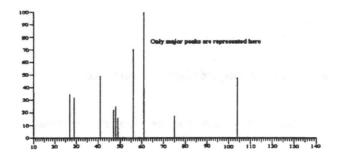

Figure 36. MS P-6.

MS Reference: **NIST 1294 NBS 101**
MS Peaks (Intensities): 61(100) 56(71) 41(49) 104(48) 27(35) 29(33)
    48(25) 47(23) 75(17) 49(16)
m/e for Dominant Isotopic Species: 104.07
IR Reference: μ **COB 3840**
IR Peaks [cm⁻¹]: 2940 1470 1430 1390 1300 1280 1240 1100 1090 1050
    960 900 880 790 750 730 690 650

SOLUTION

Resist the temptation to analyze every peak or chemical shift!

1. What strikes you at first glance when you eyeball spectra and spectral data?

2. Is the molecular weight or m/e odd or even?

3. Are remarkable heteroatoms present?

4. Subtract the weight of remarkable heteroatoms.

5. What are the possible candidates from the "Molecular Formula Compilation Table"?

6. Their calculated LU are:

7. Inspect spectral data and spectra using the diagnostic spectroscopic aids.

8. Subtract the apparent functional groups from the "best prospect molecular formula".

9. Tentative structures are

10. Reconcile your best candidate with available data.

11. Are there any discrepancies which you can't explain?

GIVEN: IR absorptions, MS spectrum, ^{1}H NMR chemical shifts, m/e
FIND: Identity of P-7

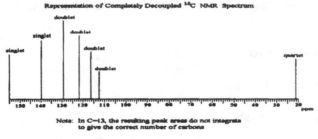

Note: In C-13, the resulting peak areas do not integrate
to give the correct number of carbons

Figure 37. ^{13}C NMR P-7.

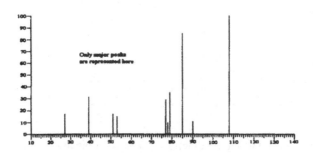

Figure 38. MS P-7.

MS Reference: **NIST 53611 WILEY 336**
MS Peaks (Intensities): 108(100) 107(85) 79(35) 39(31) 77(29) 51(17)
    27(17) 53(15) 90(11) 78(10)
m/e for Dominant Isotopic Species: 108.06
IR Reference: **COB 4818**
IR Peaks [cm^{-1}]: 3350 3040 2920 1610 1590 1490 1460 1310 1280 1270
    1230 1150 1080 1040 1000 930 880 850 780 740 690
UV Reference: **SAD 622**
UV Peaks [nm] (Absorp. Coef.): 278 273 MeOH
^{13}C NMR Reference: **JJ 247**
^{13}C NMR Shifts [ppm]: FT 21.1 112.5 116.2 121.8 129.4 139.8 155.0
    CDCl$_3$
^{1}H NMR Reference: **VAR 160**
^{1}H NMR Shifts [ppm]: 2.3 5.7 6.9 CDCl$_3$

88

SOLUTION

Resist the temptation to analyze every peak or chemical shift!

1. What strikes you at first glance when you eyeball spectra and spectral data?

2. Is the molecular weight or m/e odd or even?

3. Are remarkable heteroatoms present?

4. Subtract the weight of remarkable heteroatoms.

5. What are the possible candidates from the "Molecular Formula Compilation Table"?

6. Their calculated LU are:

7. Inspect spectral data and spectra using the diagnostic spectroscopic aids.

8. Subtract the apparent functional groups from the "best prospect molecular formula".

9. Tentative structures are:

10. Reconcile your best candidate with available data.

11. Are there any discrepancies which you can't explain?

GIVEN: IR spectrum, MS spectrum, m/e, UV absorptions, ^{13}C chemical shift predictions

FIND: Identity of P-8

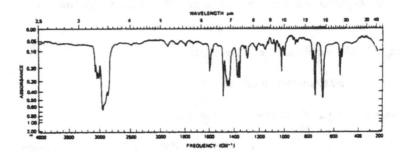

Figure 39. IR P-8. (With permission from Bio-Rad/Sadtler Division.)

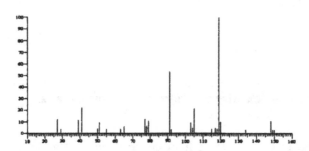

Figure 40. MS P-8.

MS Reference: **NIST 34944 NBS 441**

MS Peaks (Intensities): 119(100) 91(54) 41(22) 77(13) 27(13) 39(12) 148(11) 79(11) 120(10) 51(10)

m/e for Dominant Isotopic Species: 148.13

IR Reference: **COB 3604**

IR Peaks [cm^{-1}]: 3100 3060 3040 2970 2930 2890 1600 1500 1460 1450 1370 1360 1300 1090 1040 1000 790 760 700

UV Reference: **OES 1-409**

UV Peaks [nm] (Absorp. Coef.): 268(110) 264(151) 261(158) 258(195) 253(155) 248(123) iso

^{13}C NMR Reference: **ChemWindow Carbon-13 NMR Module**

^{13}C NMR Shift Prediction [ppm]:8.6 29.0 37.7 38.0 125.2 125.4 128.1 147.1

90

SOLUTION

Resist the temptation to analyze every peak or chemical shift!

1. What strikes you at first glance when you eyeball spectra and spectral data?

2. Is the molecular weight or m/e odd or even?

3. Are remarkable heteroatoms present?

4. Subtract the weight of remarkable heteroatoms.

5. What are the possible candidates from the "Molecular Formula Compilation Table"?

6. Their calculated LU are:

7. Inspect spectral data and spectra using the diagnostic spectroscopic aids.

8. Subtract the apparent functional groups from the "best prospect molecular formula".

9. Tentative structures are:

10. Reconcile your best candidate with available data.

11. Are there any discrepancies which you can't explain?

GIVEN: IR spectrum, MS spectrum, m/e, UV absorptions, ^{1}H NMR and ^{13}C NMR chemical shifts

FIND: Identity of P-9

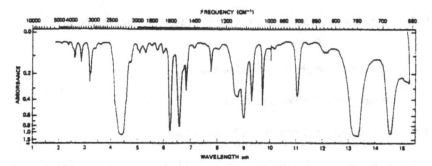

Figure 41. IR P-9. (With permission from Bio-Rad/Sadtler Division.)

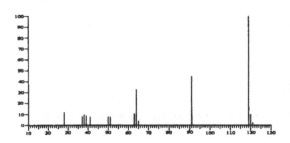

Figure 42. MS P-9.

MS Reference: **NIST 20462**

MS Peaks (Intensities): 119(100) 91(45) 64(33) 28(12) 63(11) 120(10) 38(10) 39(9) 51(7) 41(7)

m/e for Dominant Isotopic Species: 119.04

IR Reference: **COB 4344**

IR Peaks [cm^{-1}]: 3080 2280 1600 1510 1110 750 690

UV Reference: **OES 4-100**

UV Peaks [nm] (Absorp. Coef.): 277(468) 270(575) 263(457) 256(389) 226(10965) hx

^{13}C NMR Reference: **JJ 228**

^{13}C NMR Shifts [ppm]: FT 124.7 125.7 129.5 133.6 CDCl$_3$

^{1}H NMR Reference: **SAD 80**

^{1}H NMR Shifts [ppm]: 7.1 CDCl$_3$

SOLUTION

Resist the temptation to analyze every peak or chemical shift!

1. What strikes you at first glance when you eyeball spectra and spectral data?

2. Is the molecular weight or m/e odd or even?

3. Are remarkable heteroatoms present?

4. Subtract the weight of remarkable heteroatoms.

5. What are the possible candidates from the "Molecular Formula Compilation Table"?

6. Their calculated LU are:

7. Inspect spectral data and spectra using the diagnostic spectroscopic aids.

8. Subtract the apparent functional groups from the "best prospect molecular formula".

9. Tentative structures are:

10. Reconcile your best candidate with available data.

11. Are there any discrepancies which you can't explain?

GIVEN: IR spectrum, ^{1}H NMR spectrum, MS peaks and respective intensities, m/e, UV absorptions
FIND: Identity of P-10

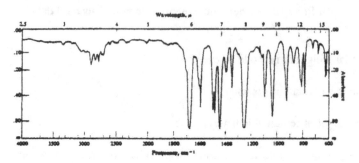

Figure 43. IR P-10. (With permission from Bio-Rad/Sadtler Division.)

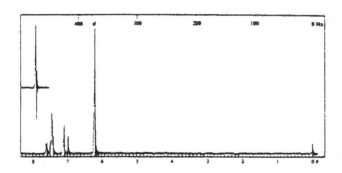

Figure 44. ^{1}H NMR P-10. (With permission from Bio-Rad/Sadtler Division.)

MS Reference: **NIST 65142**
MS Peaks (Intensities): **149(100) 150(78) 63(33) 121(27) 65(20) 38(14) 62(13) 91(12) 39(11) 61(10)**
m/e for Dominant Isotopic Species: 150.03
IR Reference: **SADG 15466**
IR Peaks [cm^{-1}]: 3030 2857 1695 1613 1471 1449 1389 1351 1250 1075 1020 926 870 800 775
UV Reference: **SAD 1178**
UV Peaks [nm] (Absorp. Coef.): 312(8630) 272(7060) 231(16600) MeOH
^{1}H NMR Reference: **VAR 187**
^{1}H NMR Shifts [ppm]: 6.1 7.0 7.4 7.5 9.8 CDCl$_3$

94

SOLUTION

Resist the temptation to analyze every peak or chemical shift!

1. What strikes you at first glance when you eyeball spectra and spectral data?

2. Is the molecular weight or m/e odd or even?

3. Are remarkable heteroatoms present?

4. Subtract the weight of remarkable heteroatoms.

5. What are the possible candidates from the "Molecular Formula Compilation Table"?

6. Their calculated LU are:

7. Inspect spectral data and spectra using the diagnostic spectroscopic aids.

8. Subtract the apparent functional groups from the "best prospect molecular formula".

9. Tentative structures are:

10. Reconcile your best candidate with available data.

11. Are there any discrepancies which you can't explain?

GIVEN: MS peaks and respective intensities, m/e, IR spectrum, ^{1}H NMR spectrum, UV absorptions, ^{13}C NMR chemical shifts
FIND: Identity of P-11

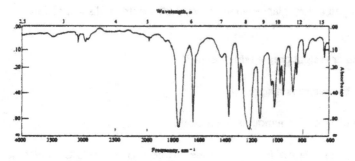

Figure 45. IR P-11. (With permission from Bio-Rad/Sadtler Division.)

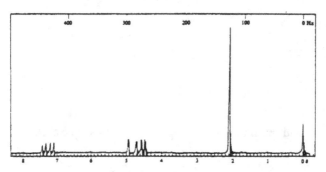

Figure 46. ^{1}H NMR P-11. (With permission from Bio-Rad/Sadtler Division.)

MS Reference: **NIST 27875 WILEY 158**

MS Peaks (Intensities): 43(100) 28(45) 42(26) 44(24) 86(20) 31(10) 32(7) 29(7) 45(2) 41(2)

m/e for Dominant Isotopic Species: 86.04

IR Reference: **SADG 8112**

IR Peaks [cm^{-1}]: 3100 2980 1760 1650 1470 1290 1220 1130 1020 970 950 880 850 790

UV Reference: **OES 1-29**

UV Peaks [nm] (Absorp. Coef.): 258(1) hx

^{13}C NMR Reference: **JJ 61**

^{13}C NMR Shifts [ppm]: cw 20.2 96.8 141.8 167.6 diox

^{1}H NMR Reference: **VAR 65**

^{1}H NMR Shifts [ppm]: 2.1 4.6 4.9 7.3 CDCl$_3$

SOLUTION

Resist the temptation to analyze every peak or chemical shift!

1. What strikes you at first glance when you eyeball spectra and spectral data?

2. Is the molecular weight or m/e odd or even?

3. Are remarkable heteroatoms present?

4. Subtract the weight of remarkable heteroatoms.

5. What are the possible candidates from the "Molecular Formula Compilation Table"?

6. Their calculated LU are:

7. Inspect spectral data and spectra using the diagnostic spectroscopic aids.

8. Subtract the apparent functional groups from the "best prospect molecular formula".

9. Tentative structures are:

10. Reconcile your best candidate with available data.

11. Are there any discrepancies which you can't explain?

GIVEN: IR spectrum, MS spectrum, m/e, ^{1}H NMR and ^{13}C NMR chemical shifts

FIND: Identity of P-12

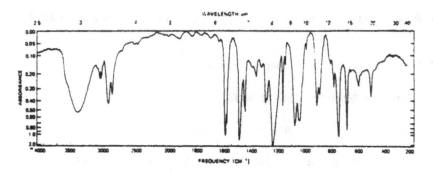

Figure 47. IR P-12. (With permission from Bio-Rad/Sadtler Division.)

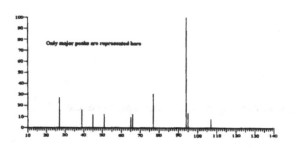

Figure 48. MS P-12.

MS Reference: **NIST 21124 WILEY 1240**

MS Peaks (Intensities): 94(100) 77(31) 51(16) 39(16) 95(14) 66(13) 45(13) 65(10) 27(10) 107(7)

m/e for Dominant Isotopic Species: 138.07

IR Reference: μ **COB 1055**

IR Peaks [cm^{-1}]: 3450 2940 2860 1610 1490 1450 1300 1240 1180 1090 1040 920 890 790 750 690

UV Reference: **SAD 99**

UV Peaks [nm] (Absorp. Coef.): 277(1420) 271(1690) 220(7740) MeOH

^{1}H NMR Reference: **VAR 506**

^{1}H NMR Shifts [ppm]: 2.7 4.0 7.2 CDCl$_3$

SOLUTION

Resist the temptation to analyze every peak or chemical shift!

1. What strikes you at first glance when you eyeball spectra and spectral data?

2. Is the molecular weight or m/e odd or even?

3. Are remarkable heteroatoms present?

4. Subtract the weight of remarkable heteroatoms.

5. What are the possible candidates from the "Molecular Formula Compilation Table"?

6. Their calculated LU are:

7. Inspect spectral data and spectra using the diagnostic spectroscopic aids.

8. Subtract the apparent functional groups from the "best prospect molecular formula".

9. Tentative structures are:

10. Reconcile your best candidate with available data.

11. Are there any discrepancies which you can't explain?

GIVEN: MS peaks and respective intensities, m/e, IR spectrum, ^{1}H NMR spectrum, UV absorptions, ^{13}C NMR chemical shifts

FIND: Identity of P-13

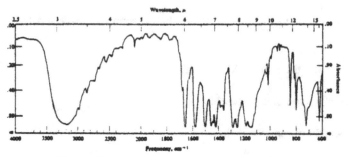

Figure 49. IR P-13. (With permission from Bio-Rad/Sadtler Division.)

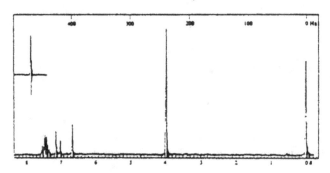

Figure 50. ^{1}H NMR P-13. (With permission from Bio-Rad/Sadtler Division.)

MS Reference: **NIST 4909**

MS Peaks (Intensities): 152(100) 151(97) 81(30) 109(21) 51(19) 52(18) 53(17) 123(16) 29(12) 153(11)

m/e for Dominant Isotopic Species: 152.05

IR Reference: μ **COB 2530**

IR Peaks [cm^{-1}]: 3330 2940 1670 1590 1520 1450 1430 1370 1300 1270 1210 1150 1120 1030 820 780 730

UV Reference: **OES 3-153**

UV Peaks [nm] (Absorp. Coef.): 309(10471) 279(10233) 232(14454) 208(12023) EtOH 301(9333) 273(12303) 228(16596) 206(13804) eth

^{13}C NMR Reference: **JJ 292**

^{13}C NMR Shifts [ppm]: FT 56.0 109.4 114.8 127.4 129.5 147.5 152.3 191.3

^{1}H NMR Reference: **VAR 197**

^{1}H NMR Shifts [ppm]: 3.9 6.5 7.0 7.4 9.8 CDCl$_3$

SOLUTION

Resist the temptation to analyze every peak or chemical shift!

1. What strikes you at first glance when you eyeball spectra and spectral data?

2. Is the molecular weight or m/e odd or even?

3. Are remarkable heteroatoms present?

4. Subtract the weight of remarkable heteroatoms.

5. What are the possible candidates from the "Molecular Formula Compilation Table"?

6. Their calculated LU are:

7. Inspect spectral data and spectra using the diagnostic spectroscopic aids.

8. Subtract the apparent functional groups from the "best prospect molecular formula".

9. Tentative structures are:

10. Reconcile your best candidate with available data.

11. Are there any discrepancies which you can't explain?

GIVEN: MS spectrum, m/e, ^{13}C NMR spectrum, IR peaks
FIND: Identity of P-14

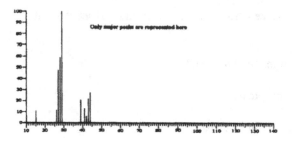

Figure 51. MS P-14.

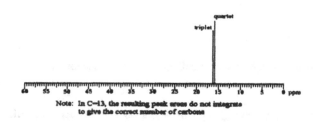

Figure 52. ^{13}C NMR P-14.

MS Reference: **NIST 61297**
MS Peaks (Intensities): 29(100) 28(59) 27(47) 44(27) 43(23) 39(21) 41(14)
  26(12) 15(11) 42(6)
m/e for Dominant Isotopic Species: 44.06
IR Reference: μ **SADP 6404**
IR Peaks [cm⁻¹]: 2940 1470 1390 1160 1060 920 750
^{13}C NMR Reference: **STOT 56**
^{13}C NMR Shifts [ppm]: 15.6 16.1

SOLUTION

Resist the temptation to analyze every peak or chemical shift!

1. What strikes you at first glance when you eyeball spectra and spectral data?

2. Is the molecular weight or m/e odd or even?

3. Are remarkable heteroatoms present?

4. Subtract the weight of remarkable heteroatoms.

5. What are the possible candidates from the "Molecular Formula Compilation Table"?

6. Their calculated LU are:

7. Inspect spectral data and spectra using the diagnostic spectroscopic aids.

8. Subtract the apparent functional groups from the "best prospect molecular formula".

9. Tentative structures are:

10. Reconcile your best candidate with available data.

11. Are there any discrepancies which you can't explain?

103

GIVEN: IR spectrum, MS spectrum, m/e, ¹H NMR and ¹³C NMR chemical shifts/predictions

FIND: Identity of P-15

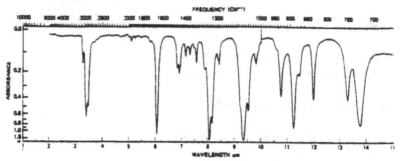

Figure 53. IR P-15. (With permission from Bio-Rad/Sadtler Division.)

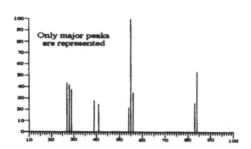

Figure 54. MS P-15.

MS Reference: **NIST 19344**

MS Peaks (Intensities): 55(100) 84(54) 27(44) 28(42) 29(38) 56(31) 39(28) 83(26) 41(25) 54(23)

m/e for Dominant Isotopic Species: 84.06

IR Reference: μ **SADP 381**

IR Peaks [cm⁻¹]: 3030 2940 1670 1450 1250 1240 1080 1050 930 890 830 750 730

UV Reference: **OES 3-43**

UV Peaks [nm] (Absorp. Coef.): 195(3802) hx

¹H NMR Reference: **VAR 111**

¹H NMR Shifts [ppm]: 1.9 4.0 4.7 6.4 CDCl₃

¹³C NMR Reference: **ChemWindow Carbon-13 NMR Module**

¹³C NMR Shift Prediction [ppm]: 29.5 32.9 72.1 97.6 142.6

SOLUTION

Resist the temptation to analyze every peak or chemical shift!

1. What strikes you at first glance when you eyeball spectra and spectral data?

2. Is the molecular weight or m/e odd or even?

3. Are remarkable heteroatoms present?

4. Subtract the weight of remarkable heteroatoms.

5. What are the possible candidates from the "Molecular Formula Compilation Table"?

6. Their calculated LU are:

7. Inspect spectral data and spectra using the diagnostic spectroscopic aids.

8. Subtract the apparent functional groups from the "best prospect molecular formula".

9. Tentative structures are:

10. Reconcile your best candidate with available data.

11. Are there any discrepancies which you can't explain?

Appendices

Bibliography

Bruno, T.J. and Svoronos, P.D.N., *Handbook of Basic Tables for Chemical Analysis*, CRC Press, Inc., Boca Raton, Florida, 1989.

Carey, F.A., *Organic Chemistry, Second Edition*, McGraw-Hill, Inc., New York, 1992.

Dyer, J.R., *Applications of Absorption Spectroscopy of Organic Compounds*, Prentice-Hall, Inc, Englewood Cliffs, New Jersey, 1965.

Lide, D.R., *Handbook of Chemistry and Physics, 74th Edition*, CRC Press, Inc., Boca Raton, Florida, 1993.

Lide, D.R. and Grasselli, J., *Properties of Organic Compounds, Version 3.10*, CRC Press, Inc., Boca Raton, Florida, 1993.

Morrison, R.T. and Boyd, R.N., *Organic Chemistry, Third Edition*, Allyn and Bacon, Inc., Boston, Massachusetts, 1975.

Pavia, D.L., Lampman, G.M., and Kriz, G.S., Jr., *Introduction to Organic Techniques, Second Edition*, Saunders College Publishing, Philadelphia, Pennsylvania, 1982.

Pretsch, E. and Fürst, A., *ChemWindow Carbon-13 NMR Module*, SoftShell International, Ltd.,Grand Junction, Colorado, 1994.

Shriner, R.L., Fuson, R.C., and Curtin, D.Y., *The Systematic Identification of Organic Compounds, Fifth Edition*, John Wiley & Sons, New York, 1964.

Silverstein, R.M., Bassler, G.C., and Morrill, T.C., *Spectrometric Identification of Organic Compounds, Fourth Edition*, John Wiley & Sons, New York, 1981.

Svoronos, P. and Sarlo, E., *Organic Chemistry Laboratory Manual*, Wm. C. Brown Publishers, Dubuque, Iowa, 1994.

Field Names

The field names in the *Guide* were taken from the *Handbook of Data on Organic Compounds* which contains over 27,000 compounds. Physical properties, spectral data, and frequently used identifiers such as the CAS Registry Number provide the user an effective "first point of reference".

CAS Index Name: Generally, the Index Name from the *8th* or *9th Collective Index* of Chemical Abstracts Service (CAS).

Molecular Formula: The molecular formula written in the Hill Order.

Line Formula: A linear array of the atoms or groups in the sequence in which they appear in the molecule.

Beilstein Reference: Citation to the 4th edition of *Beilstein Handbook of Organic Chemistry*. An entry of $3^4,250$, for example, indicates that the compound may be found in the 4th Supplement to Volume 3, on page 250.

CAS Registry Number: The Chemical Abstracts Registry Number assigned by CAS as a unique identifier for the compound.

Molecular Weight: Molecular weights are based on the 1989 IUPAC atomic weights.

Melting Point [°C]: Presence of the letters "dec" indicates the compound decomposes at the temperature indicated.

Boiling Point [°C]: A number displayed without a superscript is the normal boiling point in °C (at a pressure of 101,325 Pa or 760 mmHg). When a superscript is present, it indicates the pressure in mmHg to which the boiling point refers.

Density [g/cm³]: The temperature in °C at which the density value applies is given as a superscript.

Refractive Index (n): Refractive index is the ratio of the speed of electromagnetic radiation in free space to the speed of the radiation in another medium. The numerical value for the index may be followed by a subscript indicating the wavelength of light used and/or a superscript denoting the temperature at the time of the measurement. The absence of the subscript implies the measurement was determined using yellow light (the sodium D line).

Solubility: A relative scale of solubility is used:

      1 = insoluble,
      2 = slightly soluble,
      3 = soluble,
      4 = very soluble,
      5 = miscible,
      6 = decomposes.

m/e for Dominant Isotopic Species: This is the mass to charge ratio for the single-charged ion of the most abundant isotopic species.

Mass Spectrum [Intensities]: The most abundant peaks are shown. The relative intensities are given in parentheses, with the strongest peak assigned an intensity of (100).

Infrared [cm^{-1}]: All absorption bands characteristic of a functional group were coded. In addition, at least one strong band in each micrometer or 100 cm^{-1} interval was coded. Data originally coded in micrometers were converted to wave numbers.

Ultraviolet [nm] (Absorp. Coef.): Wavelengths of all major bands, their molar absorption coefficients (given in parentheses), and the solvent used are provided. The wavelength range from 170 to 600 nm was coded. When the spectrum showed vibrational fine structure, only the peak centers were listed.

Proton Nuclear Magnetic Resonance [ppm]: The proton chemical shifts, in ppm (δ), for specific protons were coded to ±0.1 ppm over the range 0—15 ppm referenced to tetramethylsilane (TMS). When complex spectra due to second order effects or overlapping resonances were encountered, the range was recorded. The solvent is identified.

^{13}C Nuclear Magnetic Resonance [ppm]: The carbon chemical shifts, in ppm (δ), for specific carbons were coded to ±0.1 ppm over the range 0—200 ppm referenced to tetramethylsilane (TMS). The solvent in which the spectrum was obtained is stated.

^{13}C Chemical Shift Prediction [ppm]: The chemical shift was determined by Softshell's ^{13}C NMR Module Software.

Abbreviations

ace	acetone
al	alcohol (generally means ethyl alcohol)
ALDN	Aldermaston, *Eight Peak Index of Mass Spectra*, U.K.
bz, Bz	benzene
CAS	Chemical Abstracts Service
chl, Chl	chloroform
COB	Coblentz Society spectral collection
ctc	carbon tetrachloride
cw	continuous wave
cyhex	cyclohexane
diox	dioxane
eth	ethanol
FT	Fourier transform
HBCP	*Handbook of Chemistry and Physics*, CRC Press, Inc., Boca Raton, FL
hx	hexane
iso	isooctane
JJ	Johnson and Jankowski, *Carbon-13 NMR Spectra*, John Wiley & Sons, New York
mcl	monoclinic
NBS	NBS-EPA-NIH Mass Spectral Database, NSRDS-NBS-63
NIST	NIST/EPA/MSDC Mass Spectral Database, 1990 Version
OES	Phillip et al., *Organic Electronic Spectral Data*, John Wiley & Sons, New York
os	organic solvent
SAD	Sadtler Research Laboratories spectral collection
SADG	Sadtler Research Laboratories IR grating collection
SADP	Sadtler Research Laboratories IR prism collection
STOT	Stothers, *Carbon-13 NMR Spectroscopy*, Academic Press, New York
VAR	Varian Associates NMR spectra collection
WILEY	*Atlas of Mass Spectral Data*, John Wiley & Sons, New York

Indices

CAS Registry Number Index

Compounds are listed in numerical order of their CAS Registry Number. A primary name accompanies the CAS Registry Number identifier.

Name/Synonym Index

Chemical Abstract Service names, IUPAC names, common names, and other synonyms are listed together. The order is alphabetical.

Name/Synonym Index (continued)

Answers to Practice Problems

Answers to Practice Problems

P-1   1-hexyne

P-2   1,2,3-trichloropropane

P-3   ionone

P-4   2-methyl 2-butene

P-5   *o*-nitrotoluene

P-6   butylmethylsulfide

P-7   3-methyl phenol

P-8   *t*-pentyl benzene

P-9   phenyl isocyanate

P-10   piperonal

P-11   vinyl acetate

P-12   phenoxy ethanol

P-13   *p*-vanillin

P-14   propane

P-15   3,4-dihydro-2H-pyran

Printed in the United States
by Baker & Taylor Publisher Services